ENDEAVOR

QUARTERLY RESEARCH JOURNAL — *Autumn 2020*

FROM THE STONE-AGE TO THE SPACE-AGE

FOUNDATION FOR RESEARCH OF THE ENZMANN ARCHIVE
ENDEAVOR, the Quarterly Archive Research Journal • Volume Two, #2, • Autumn 2020

Joanna Enzmann, Pres. & Treas.; Michelle Snyder Founder & VP.; Kimberly Beals, Secretary
Board of Directors: Edwin Pangman, John Peterson, Donald Crookes, Holly Snyder
Lead Editor: Michelle Snyder; Research Editors: Jay R. Snyder, Charles T. Moses, Blair March, Edmund Devine, Peyton Beals

ENDEAVOR is a quarterly research journal of Dr. and Mrs. Enzmann's life-long development of World Chronology, Geology, Planetology, Space Mission Planning, and Starship design. All proceeds benefit the FOUNDATION FOR RESEARCH OF THE ENZMANN ARCHIVE, Inc.
FREA, Inc., Publisher

ENDEAVOR #7
ISBN: 9798597952291
ENDEAVOR The Enzmann Archive Research Quarterly
All Rights Reserved, Copyright © FREA, Inc. 2021
Published in the USA by FREA, Inc.
Printed in the USA
For information: freafoundation@gmail.com

ENDEAVOR AUTUMN 2020

ENDEAVOR
QUARTERLY RESEARCH JOURNAL   *AUTUMN 2020*

FROM THE STONE-AGE TO THE SPACE-AGE

"A crucial requirement of ameliorating global problems is the projection of new, positive images of desirable, attainable futures." - *RDE*

AUTUMN 2020 ENDEAVOR CONTENTS

In Memoriam, Dr. Robert Duncan-Enzmann 1

From the FREA President .... Page 11
From the FREA Editor .... Page 12
Enzmann Starships Now .... Page 17
    Starflight: Dreams to Reality, *Dr. Tom Jones*
    Theory of Relevance, *Dr. Enzmann*
    Intelligence, Educating the Gifted, *Dr. Enzmann*
Enzmann Conference Articles .... Page 21
    Forward, Conference One, *I. B. Laskowitz*
    Preface, Conference One, *Dr. Enzmann*
    Some Implications of Extrasolar Intelligence, *Frederick Ordway, III*

Enzmann Planetology .... Page 36
    Universal Order Theory, *Dr. and Joanna Enzmann*
    Mensuration, *Dr. Enzmann*
    Rocks in My Head, *Peyton Beals*

Enzmann Cosmology .... Page 40
    Science and the Holy Scripture, *Dr. Enzmann & Michelle Snyder*
    On Evil, *Dr. Enzmann*
    Cosmological Thoughts, *Blair March*

Enzmann Chronology .... Page 46
    Ice Age Language, Mother and Child, *Dr. Enzmann & J R Snyder*
    The Immortal Child, *Dr. and Joanna Enzmann*

Symbology .... Page 60
    King Midas Unlocked, *Symbologist Michelle Snyder*

ROBERT DUNCAN-ENZMANN

in memoriam
1924 - 2020

Dr. Robert Duncan-Enzmann, In Memoriam

Public Obituary

Dr. Robert Duncan-Enzmann was born in Peking, China, in 1924 to an American mother and an Austrian father. He passed away on October 19, 2020, and is survived by his wife, six children, and fourteen grandchildren. He was preceded in death by his beloved sister, Jane. He served in the Navy during WW II as a pilot and navigator, receiving a Purple Heart. He was discharged with honor.

Dr. Enzmann came to the USA for the first time at the age of five. He grew up in Massachusetts and Maine. He earned doctorates in medicine and geology and attained three master's degrees. He conducted geological work all over the world, especially in Southwest Africa. He authored several books; his translations of ice age inscriptions were published in 2013. Dr. Enzmann is best known for his work in the pre-NASA space and military weapons industry, working with Drs. Goddard, Bussard, and von Braun, to lay the foundation for landing men on the Moon. He worked for Beryllium Corp, Avco, and Raytheon and taught at several Boston universities.

A Great Man Is Gone
FREA staff

On Tuesday, October 19, 2020, the world lost one of the most intelligent minds to have lived in centuries. Dr. Enzmann is now preserved in Suspended Animation at a cryogenics facility in Detroit. One day, he told us, he will be cloned or even revived, providing that humanity has defeated old age. At least this was his hope in life.

Robert Duncan-Enzmann was born in Peking, China, at a time just before the city's electrification, before the introduction of the motorcar. His father, Ernst von Enzmann, was an officer in Franz Josef's army. He escaped from prison in Siberia by walking to China; his mother, Florence Goodman, was a native of Maine, USA. As a Johns Hopkins graduate, she was on the Peking Union Medical College staff. Ernst and Florence met there and married.

Robert attended British Embassy schools to learn reading, writing, composition, arithmetic (emphasizing mental computation), history, astronomy, and navigationally based geography. The school regularly exchanged students with France and Germany, beginning with Kindergarten.

He has achieved degrees in engineering, physics, medicine, and law and has geological experience on every continent. He is fluent in many languages, including English, Chinese, German, Arabic, and French. He can also read several ancient languages. Dr. Robert Duncan-Enzmann's extensive travel, education, and knowledge of languages were of great benefit to his life-long effort translating ice age inscriptions, published in 2013. His accomplishments in planetology and space mission planning are too numerous to list.

Doc E lived almost one hundred years. He witnessed a century of technological advance, from life without electricity to landing men on the Moon. He saw the light bulb's invention, the radio, television, cars, computers, missiles, rockets, and space shuttles. He worked with von Braun to develop the most powerful engine ever devised – the Saturn Five.

What is best known in Doc E's portfolio is perhaps the Enzmann Starship. An interstellar ship based on a design that Dr. Bussard and Dr. Enzmann worked on – the Orion – Dr. Enzmann's starship advanced Orion's capabilities to that of interstellar travel. Yet, what is not as well known about Dr. Enzmann is the vast area of influence his knowledge had on our world. In some circles, he was known as the Missile Lord, as his work in weapons and radar has resulted in an unparalleled military capability to defend our nation.

One of the most outstanding achievements was developing the APChE (Automated Program Checkout Equipment) program for prelaunch. It was first applied to the Patriot Missile and is now used to launch everything that leaves the planet. FREA published an article about this in ENDEAVOR (spring 2019 premier issue, page 12).

Another area of knowledge that will continue to enrich humanity is his expertise in history. His magnum opus, which FREA intends to complete for him, combines all known history timelines into one. From astronomical events to zoological evolution, Doc E has notes, lists, timelines, images, and documents that must be merged into one manuscript. He explained that this is the only way to correct accounts of history contradicted by other accounts and fill in the gaps in our knowledge of Earth's complete history.

From astronomical and geological events to weapons and navigation, the histories of dozens of subjects have all been researched, and his detailed notes are abundant. Not only will FREA compile and publish the manuscript, but we have invested in an online timeline that allows entry of events over millions of years, yet in detail of day and time when needed.

Included in the History Archive are the extensive documents generated by his work translating ice age inscriptions. Like the Rosetta Stone, these translations have opened our ability to read the pictorial language of 12,500 BC and those of older and more recent. The same language style was used in 16,000 BC at Altamira and is the foundation of much more recent languages. *Ice Age Language, translations, vocabulary, and grammar,* his first book on the subject, was published in 2013.

Doc E earned a Doctorate in geology from Upsala University. He did fieldwork on every continent, mostly southwest Africa. This foundation made him valuable to the military and space tech industries, and he spent many years researching and creating beryllium, a substance with specific uses in both areas. FREA's Museum Archive has many samples of his efforts to make this substance. Doc E understood that geology was an essential area of knowledge to explore planets and asteroids. So why did the USA wait until the last Moon landing to send a geologist?

In the Rhodes Fairbridge Encyclopedia of Geomorphology, Doc E published his Expanded Order Theory. It is a fundamental subject of study for those interested in Cosmology.

Dr. Enzmann's medical doctorate benefitted from his geological knowledge, especially of hexagrams' and pentagrams' natural formation, which he said were valuable in diagnostic observation. This diagnostic discovery was the basis of his thesis, which FREA hopes to publish.

Doc E started writing at age five and has done so ever since. Educated in RAJ schools in China and Germany, he had many unusual interests for a young boy, including rocket fuel. Taught math calculations done mentally, languages, and history, he gained an education by teachers in their 80s, who were also taught by teachers in their 80s, providing an education covering two centuries of personal experience as well as the academic subjects. He learned King's English in the RAJ schools and Chinese on the streets of Peking. By adulthood, Doc E could read and or write over a dozen languages.

Doc E also told stories. Many stories. Some were about history, told as verisimilitudes, and some were technological sci fi. Some were about his life. Without these stories, the Enzmann Universe would not exist.

His stories grew from the reality of his knowledge, his work, and his futurist vision. RealScience Fiction is based on hard, workable, feasible science. Doc E wants readers to immerse themselves in a possible world of starships, and his stories tell us what kind of life that would be. We offer these stories here, in the Enzmann Chronicles, for your enjoyment and inspiration. Starships could have been a reality in our current century had the powers of government allowed the Grand Design to manifest. Alas, they did not.

In story and real life, Dr. Robert Duncan-Enzmann and Joanna Enzmann changed our country's face and altered the world's status quo. Their work in space engineering, missile defense, radar, and computer programming is the foundation of every rocket and missile that leaves the ground and every protection against airborne attack. It is the foundation on which the USA sent men to the Moon.

Here is a brief list of his educational and professional accomplishments:

British Embassy School, Peking, China; WW II USNAC; AB Harvard; ScB Hon., London; Standard, MSc, Witwatersrand; Nat Sci Scholar; MIT course work; Royal Inst. Uppsala Swed.; Ph.D./MD Cuidad Juarez, Mex.; Pacific Radar: Greenland Gap-filler, Canada DEW-line; SAGE; Pacific PRESS; California ATLAS, BMEWS; ICBM; Kwajalein Atoll ICBM intercept; TRADEX; Mars Voyager; Cryptography.

At FREA, our mission is to publish the vast amount of writing and imagery Doc E produced in his more than nine decades of life. The Archive contains his stories, his sciences and history, translations of ice age inscriptions, timelines, cosmology, astronomy, medicine, geology, radar, physics, mathematics, planetology, engineering, and archaeology, and more.

The Archive also includes objects from his life, travel, and work on every continent. FREA is founding a museum and gallery as part of the Enzmann Legacy.

His widow asks that in place of flowers, memorial donations be sent to the nonprofit that publishes Dr. Enzmann's work: Foundation for Research of the Enzmann Archive, Inc.

Donations can be sent online via PayPal to freafoundation@gmail.com. Checks can be mailed to 70 North Street, Grafton, MA 01519.

ENDEAVOR

FROM THE STONE-AGE TO THE SPACE-AGE

Robert Duncan Enzmann on Deck, Enzmann Archive Image

Amazon Station

Robert Duncan Enzmann
A true story from Doc E's life in the Navy.

It's a frigate, no name, just a number. 72nd was built shortly after WWI, the war to end wars, and named the Patrick Henry. It was then completely modernized just before WW II, the Good War, and renamed 721. Yes, the crew endlessly chips off paint and then re-paints, but some parts of the old name show through the new paint.

I'm aboard accidentally; I was with a group waiting for something – who knows what. Likely not even the planners who by now have personally forgotten and administratively lost us forever in the tidal wave of paper engulfing them.

Because of the war, the old man, the Captain, will be in the Navy way over thirty years before he retires. With him, which is unusual, are a half dozen Bos'ns, hard-bitten, tough guys and highly competent seamen, all utterly loyal to the old man.

Captain: "This counter-current isn't ephemeral."

"What, Sir!"

"Doesn't come and go; it is always somehow there."

A Bos'n looks at me and mutters, "we see em come and see 'em go."

I look lengthily at the chart. The counter-current sketched artistically on it with numbers, show its speed and direction.

Captain to Bos'n at the wheel, referring to me: "Teach this guy how to take the wheel."

I'm an object of contempt, the target of lots of curses and sometimes sneering laughs. I just take it for the whole watch. I am awful at the wheel. I am greatly relieved at the end of the watch; I go and have lunch at the general mess dining region just off the galley, then return to the bridge.

Bos'n: "What the hell do you want?"

Risking getting beat up, I say, "Asshole! Obey the Captain's orders. No one learns on one watch, so teach me."

I'm awful, get cussed out, shoved in, shoved out. The Bos'ns are pissed.

The least physically powerful watch is a 12 pm to 4 am. On watch, the Bos'n won't risk a fight.

He's alone, so I say, "Prick! I know lots of things and know them well. You don't know much; you ain't got the brain. But, and I hate to admit it, what you do know, you know very well. Follow your orders, teach me everything you know."

He says: "My God! You're back!"

I say: "The captain is sleeping, so follow orders."

His relief takes the wheel: "They come, and they go. But this 'fuckface' is stupidly stubborn."

This battle continues for some weeks, but I learn $+-25^0$, $+- 10^0$, $+-$ to 2^0. Then suddenly, I am as good as any of them, then better than all but one.

Chief Bos'n: "Captain, when do we get rid of Fuckface?"

Captain: "We just might not. Give him a few more really tough turns, see if he's got the stuff."

I have it out with the Chief Bos'n in a corridor. He reaches to shove me, I am very quick, trip him, and he is down.

Bos'n Mac: "When I get up, I'll beat the shit out of you!"

I say: "Sure you can, but it was worth it, and as you get up I get one more sucker punch."

"Worth it!" exclaims Chief Bos'n and laughs.

I give him a hand up, and we both laugh.

The old man is a great and willing teacher. I now stand a daily watch at the helm and soon stand-in for the Captain. The Bos'ns, whether the captain is or isn't there, answer me with 'yes sir.' I don't quite realize it as it develops over weeks, then months; how it has all changed. I am no longer just being taught or even tutored by the old man. We now converse as equals.

The torpedo hits near the bow with a terrific crash. In my cabin, I'm slammed down, hitting the hinged-out table. My head's bloody, and I stagger about on deck mumbling, "head cuts bleed too much."

The torpedo was running near the surface. The bridge is shattered. We are taking on water but not too fast, lots of time to launch our several whaleboats, which are better than WW II standard Navy lifeboats. Our ship will go down bow first but slowly, and I realize all three of our lifeboats will be both supplied and launched. Dizzy, I sit to wait and pass out, then wake up, still dizzy. My head hurts. A guy is marking on my head – I half pass out and look around. I see only one boat.

"We have three boats. Where's the other two?"

Chief Bos'n: "The ninety-day-wonder first and second mate college-boy officers took off, precision rowing, chanting as they went. We just decided to drift till you wake up. In this boat, it is us Bos'ns and old sailors."

I say: "What the hell am I doing at the tiller with you apes."

Bos'n Charlie. "Go easy on us Bob; we elected you Captain. The old man went down, was dying from the blast."

I gave them my orders: "Hoist sail, run before the wind, which will be onto the land until evening. Then drop a sea anchor and drift."

A guy stands up to protest, the Bos'n draws and aims a colt revolver saying, "SIT DOWN! Not another word or die for mutiny."

I say: "Mac! Take the tiller, you know I can't – too sick and no experience."

Bos'n Mac: "Yeah, we know, but you and the old man are the only ones that know these deadly waters in the Amazon current. It's far, far eastward out to sea."

Then I say: "Here's our situation, either we are gripped by the Argentina current of the South Atlantic gyre or caught by the Gulf of Mexico intake for a long, long blistering hot trip of many weeks. Cold death toward the South Pole, or scorching death along the equator."

Bos'n Charlie: "So?"

"I'll get us to Brazil. It can be done." I hold my head. My God, that hurts.

"Doc is fixing you, looking for concussion killing germs. He's a 2nd class pharmacist mate and is going to study and be a real pharmacist aft the war."

Dizzy, I settle down and drift in and out, mumbling, "string, fishing sinker, any watches? Compass. Stay before the wind."

"Wake me before sunset at wind hush. Can you tack?"

Bos'n Charlie: "Yes!"

"Tack eastward into the day wind."

Bos'n Mac: "I ration out water at nightfall. It lasts longer in bodies than in the daytime. Our carboys are all full; we have plenty."

"We are crossing a brown streak. Give me a cup and halazone tablet; it will dissolve, killing germs."

I demonstrated drinking the tinted water (with silt and vegetation juices). It's fresh. "Ok. Halazone in the carboys, then double the water ration."

Counting the cupful I just got, I get an extra cup full. All hands almost shout, "He deserves it."

Then I say: "Everyone put on hats. I'm going to read words from the Bible. It's he who taught me about water and everything else. Respect him, not so much me."

"At critical times we will have to row."

The four tough Bos'ns replace weaklings at the oars. A guy with a drum plays a drum song.

Tup, tup, tup, boom boom boom all together pull.

T T T boom. On and on across the first stretch of clear water.

T T T boom. It's Sammy and Ratface. Rest the oars, ride westward on the yellow loop.

T T T boom boom bend to the oars, rest oars and ride on the current.

We made it across all three countercurrents that day, gaining many tens of miles westward.

I say: "Now, set sail and ride down-wind with the Amazon night wind behind us."

Land. It's just a far-off smudge on the horizon.

Bos'n Mac: "Row?"

"Absolutely not. Tack. We will need real strength in the surf."

On this boat, very fortunately, the oarlocks are positioned such that you can push the oars straight down and brace them against the oarlocks. There's an Eckman spiral counter-current below an opposite surface-current. I am pleased with myself and the old man's months of lessons. The seamen see pure magic. Mac gives the orders after I explain.

"Hoist old glory, I know we have a flag in our stern locker. Tell them who we are. Americans. Stars and stripes unbeaten."

With that, I'm fuzzily dizzy again. The well-experienced Bos'ns bring us through the surf, beach us, and carry me onto the shore. I'm worn out and infected in my head and parts of my back. They take most of us to the local hospital.

Bos'n Mac asks: "How the hell did you throw me for a loop, and then stand there with a club, declaring 'one more sucker punch as you get up.' Then you bragged, 'it's worth it.'

You know I can beat the shit out of you. I'm triple your size."

"I waylaid you."

Bos'n Ralph: "That don't buy nothing. We beat the hell out of punks in most of the world's big ports, sometimes three or four at a time. Waylaying don't do nothing. How the hell did a twerp like you do it."

"OK, shit for brains. Here's how. At the corner in the passageway, squares of plastic the same color as the floor. Two layers of plastic with grease between. More slippery than any ice. Grabbed his arm by the sleeve, pushed with my feet. Down he went, not on his fat ass, but on his back."

I turn to Mac: "Expected you to come up roaring like a bull and had another patch ready, but full of bull you're a smart bull, surviving a bullfight."

Bos'n Mac: "The old man soon knew. He just laughed at me."

Bos'n Charlie: "We been through two big wars and a few little ones with the old man. It's him who had us work over wise-ass guys so he can see 'em go. You are the second in 23 years who lasted."

As we approach shore we talk about the other two lifeboats.

Bos'n Charlie: "It was a tragic way for the old man to die. We saw the college boy ninety-day-wonder naval officers (trained in 3 months), and our first and second mates go in the lifeboats. Yeah, they dumped on the old man and you too. We figure they're caught in the Argentine current, headed toward the Antarctic. It's a terrible way to die."

I wake up later in a hospital bed.

Bos'n Mac: "You were out like a turned off light. The USA counsel was here. We told him about the other two boats. Tragic, we are in a war. Only one PBY (amphibious aircraft) will look for them. It's because we have a war to fight and it's better to save ships. Our planes scare the kraut submarines."

The other two boats are never seen or even heard of again. It is a terrible way to go, especially for the last one, who will survive several months before his hopeless death. It is possible they might reach Saint Paul's rock and climb it to survive by eating sea birds.

Some millions of Germans, Austrians, and Galicians live in Brazil's Santa Catharine Province. They started settling there before the American 1776 revolution. Roosevelt had just begun to exterminate them by having them shipped into the Amazon's malaria-infested jungle.

There are more than a few in the hospital and many visitors. My German is as good as any of theirs. And better than most. They take me for one of theirs, which I wasn't, and never will be. Their ideas are frozen into the 1938 notion that they can't lose this war. It's 'impossible,' they brag, and again they are dead wrongly for a second time, again 'snatching defeat from the jaws of victory.' 'We can't lose' is a plausible 1938 to 1940 contention but look at Hitler. A good soldier, a so-so artist, risen to his level of incompetence, and look at his trail of wreckage.

I just say: *Anyone can lose a war. At least consider the possibility*. The discussion ends. 'Tis impossible.

Bos'n Mac: "You're *talking* with them. You are a fucking *Heinie*."

"And you, Bos'n Mac, are a fucking stupid Irish Mic."

Bos'n Charlie: "He talks French too."

Bos'n Ralph: "Fucking Heinie, his Ma was a Kelly of the stupid Irish Mics. Talks French, a slimy Frog."

"So, what we got?" sneers Bos'n Charlie at Mac, "the worst of the worst – an Irish Mic, also at the same time a slimy Frog Heinie."

I say: "Eisenhower is a Heinie, so was world war one general Black Jack Pershing, and oh my, dig into Shirley Temple's ancestry. Got problems…. got problems with that."

Bos'n Mac. "I didn't know about general Pershing."

Bos'n Ralph: "If a cat has kittens in an old oven in the yard, they are still not biscuits."

Bos'n Charlie: "Bob, it's Mac who got us to elect you as captain."

Bos'n Mac. "Through the years only two have survived old man's 'get em' on the way. 'They came,' the old man said, 'let's see 'em go.' You're the second one. The first is a three- or-four-star admiral now. You will do the same and better."

I say: "The old man also spoke French and German, and some Spanish. When I get out of here I am going into Captain Eddy's radio training program, and from there into the GCA, the ground-controlled approach program, invented by Alvarez."

Note: The 'old man' was later the discoverer of the asteroid impact on Yucatan that exterminated the dinosaurs and was also involved in taking fuzes and bomb-grade uranium from the German submarine surrendering at Portsmouth, NH, to the Manhattan Project.

Bos'n Mac: "It would be best if we all stayed together with you on a new frigate."

I say: "The bureaucrats don't and won't see it – I don't have the right background."

And I was right.

5-toed Dragon with Pearl, Joanna Enzmann

From the FREA President

Year's End
Joanna Enzmann, President, FREA, Inc.

As the year winds down toward its end, it has been a memorable one, to say the least. Most of the memories have not been good ones: from a worldwide spread of a deadly disease to being confined to quarters for most of the year, to loss of employment, to seemingly never-ending hurricanes, to the messiest Presidential election since 1876-7 (look it up, Hayes vs. Tilden).

The saddest memory of this year will be the loss of my husband of 62 years, the namesake of the Enzmann Archives. FREA will live on, expanding in scope, publishing, research, building on the amazing mountain of knowledge embodied in the archives. Even now, after a year and a half of work, sorting, cataloging, filing, etc., the work has barely begun. Endeavors have given a glimpse so far into the variety of material that exists, and more waits to be examined. I hope that the next year will see some moderating of this year's upheaval, and we can all breathe a sigh of relief.

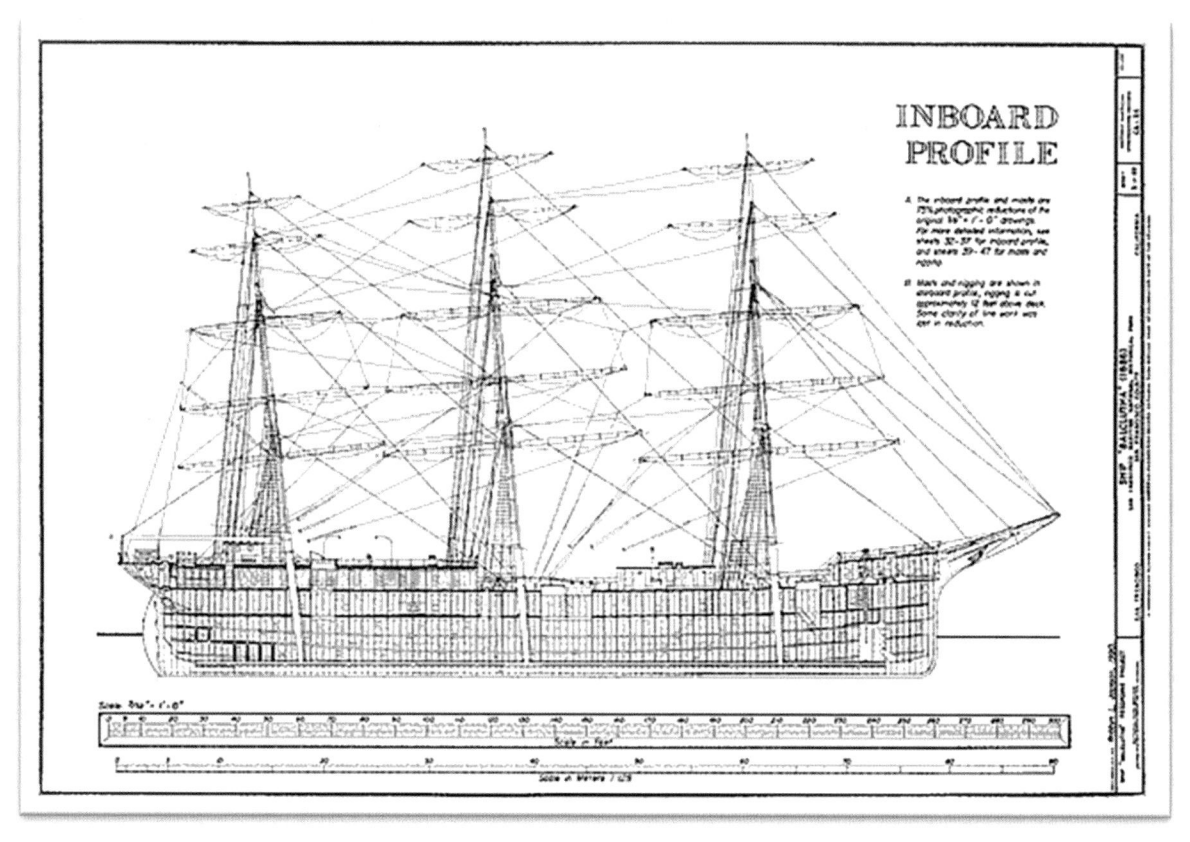

Inboard Schooner profile, Enzmann Archive Image

From the FREA Editor

FREA News
Michelle Snyder, Founder, VP, Editor

To say that 2020 has been a challenging year is an understatement. This year has, in a very historical sense, been uniquely challenging. Not since the Great War has our country faced such nationwide struggles.

Yet, with all struggle comes strength. As individuals and businesses, we must play the hand we are dealt and still try to win. Doc E taught us to stay focused on the task, never quit, and be skeptical. His passing is unique, for he is now in cryogenic stasis, awaiting advances that will allow him to return someday. He is also on every piece of paper and in every computer file FREA has that were created by his hand. We are surrounded by Dr. Robert Duncan-Enzmann.

Our mission has been and will continue to be compiling, editing, and publishing the information poured forth from one of the greatest minds to live in centuries. And yet we are aware that in the Archive is only a small percentage of what Doc E really knew. The rest sits in his brain, waiting. Ironically, his knowledge helped create the foundation on which his human hopes rested for his fate after death. He helped found the cryogenic science and business which now holds his hoped-for future.

So, we forge ahead in FREA, researching the seemingly endless paper and objects in the Archive, piecing together the life and knowledge of an enigmatic life.

In other FREA news, we are sad to lose Lisa Bengiovanni from our Board but wish her the best as she builds a new life with her fiancé Heath VerBurg, a FREA friend for more than a decade. Congrats to you both – a match even the stars cannot deny.

In her place, we welcome Holly Snyder, an experienced paralegal who is studying natural sciences. Her fiancé is the grandson of Doc E – we couldn't have asked for a better fit. Holly will also work once a week for FREA, two days with our President, and two days with the Science Archive. Another addition to our Board is Ron Haley, a Past Master in Freemasonry, an expert analyst working with AI and machine learning, and a member of the Board of Lunar Station Corporation. Ron is a natural at networking and connections. His addition to our Board is a perfect fit.

We are happy to report that FREA has taken in about $5,000 from our Tag Sale and subsequent sales. We still have in our inventory of items to sell many beautiful pieces of crystal, dolls, stamps, coins, shells, gems, silver, china, and toys and gifts. We will eventually offer these items online.

FREA has also received the proposal for architectural changes to the two buildings, expanding each one's space for added function and comfort. Significant repairs to the roof will be done at the same time. A shed dormer will be added to the back of the West Wing and the East Wing. When this is accomplished, we will move ahead with the renovations necessary for the Museum space. John Marro, our architect, has taken meticulous measurements and is a creative thinker about how to update our 300-year-old property.

At this time our newest publication, *Enzmann Chronicles*, is in its seventh month! Our editors are compiling a unique story, one of Doc E's best. Althea is about an AI Robot Tank that 'wakes up.' Written in the early 60s, it is the back story to the AI starship engineering in the Enzmann Universe. It will be published as a serial each month, along with Wagon Train to the Stars and other short features.

Like all schools, New England SciTech has faced challenges holding its classes. As life returns to something resembling normal, FREA will once again donate to the FREA/NEST scholarship, which the school will use as they see fit.

Perhaps the best news is our contract with in-Think, a business growth and marketing company. We are being discovered, rebranded, and rebranded with a new website and lead campaigns in the marketing effort. A most generous donor has covered the rather hefty cost of having this top-of-the-line company take us on. FREA is a challenge to anyone's creative thinking. We are excited to see what we look like, who our target markets are, and most especially, to begin to bring in revenue from sales of publications and donations to support the research it takes to publish.

Finally, at FREA, we hope you had a wonderful holiday season. Despite the challenges of 2020, we must all find reasons to be Thankful and practice the charity and generosity, which is key to the season of giving and having a better next year.

ENDEAVOR

FROM THE STONE-AGE TO THE SPACE-AGE

1: 1969 Nagaland – Moth

From the FREA Archivist

Stamps: Not Just for Snail-Mail Anymore
Kimberly Beals – FREA Archivist

Stamp collecting, or philately, is both fascinating and fun! Exploring a collection that was not yours and digging deep to determine why it was collected is an adventure that I never knew I would enjoy undertaking.

Why collect stamps? They have colorful images, real-life stories, various topics, depict history, and have monetary value. The value of old postage stamps varies by era and country. Different aspects of postage stamps, like watermarks and perforations, are prevalent, especially with old postage stamps. A definitive stamp is a postage stamp that is issued regularly in a particular country. Some are rare or valuable and only a philatelist, or a stamp appraiser, can distinguish the difference between a rare definitive and an ordinary postage stamp. Countries also issue commemorative stamps that can offer historical information and potential value. There are cases where stamps are printed but never issued as legal postage. These stamps are known as "Cinderellas." Even these can be of interest and are sometimes collected. The Enzmann Archive includes several examples of Cinderella issues *(see above Fig. 1)*.

In addition to the most common rectangular shape, stamps have been printed in geometric (circular, triangular, hexagonal, and pentagonal) and irregular shapes. The United States issued its first circular stamp in 2000 as a hologram of the earth. Sierra Leone and Tonga have issued stamps in the shapes of fruits.

The first stamp, which was the Penny Black stamp, issued in Great Britain, was put on sale on May 1, 1840, and was valid on May 6, 1840. Today it is valued at $5,000 if in good condition. The first United States postage stamp, for distribution throughout the country, was prepared under the authorization of the Congressional Act of March 3, 1847. Early U.S. stamps can exceed $10,000 in value. This was a good investment for a few pennies in the late 1800s.

To put an amateur view to Dr. Enzmann's Stamp collection, I first thought that the interest in collecting started with his father, Ernst von Enzmann. He had a group based on his travels. I found a wonderful album that contained many stamps dating to the late 1800s from such countries as Austria, Great Britain, and Germany (see Fig. 2).

Dr. Enzmann's collection was not as well organized as his father's but was significantly larger. Stamps from countries where he had lived, traveled, or worked commemorated NASA and space travel, geology, renowned scientists, history, and other countries' customs. The collection also included stamps depicting horses, a personal interest of his wife, Joanna Enzmann (see Fig. 3).

Many of the stamps have little or no value other than sentimental. Perhaps the attraction was in the design or theme. I started believing that only stamps in mint condition or un-canceled were of any value. It turns out that many stamps that have been canceled can have a value, depending on whether that country still exists or had changed identity over the years.

Will the introduction of the "Forever" Stamp mean that stamp collecting might one day fade away and become a hobby that future generations never know or understand? Hopefully, collections will be passed down and preserved for history, even if no significant monetary value exists.

The undertaking of sorting, researching, and curating Dr. Enzmann's stamp collection not only strained my eyesight but enlightened me in a subject that I knew very little. Some of the collection will be retained for the Enzmann Archive Museum; the rest will be sold to support the ongoing efforts of FREA.

2: 1883 Austria – Coat of Arms

3: 1967 Yemen – Arabian Horses

Stamps from the Enzmann Archive Collection

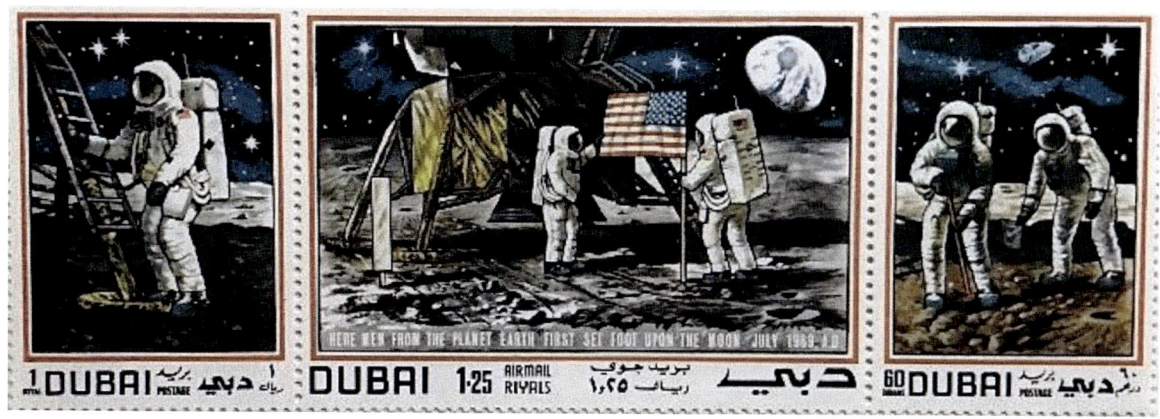

1969 Dubai - Moon Landing

1968 People's Republic of China

1970 Taiwan – 100 Horses Paintings

1969 Oman State - Parrots

Enzmann Starships Now

Starflight: Dreams to Reality
Dr. Robert Duncan-Enzmann
Dr. Tom Jones

To support life for lengthy periods, adequate supplies of air, water, and food must be carried, recycled, or produced. For instance, water would be recycled, and the raising of crops would constitute a major activity. Waste disposal would necessitate extensive recycling. Other needs that would be met include privacy, quiet, and open spaces. Human needs' systematic satisfaction is routine at remote sites in Antarctica and aboard nuclear submarines, cruise ships, and space stations. The Echolance is designed to meet such needs in ways conducive to enhancing the quality of life.

Protecting voyagers from the reactors would be much easier than shielding them and the ship from disastrous collisions with particles. Even protons in space would constitute hazards to people moving at speeds approaching that of light. The most effective and certainly the most unique of the many types of shields that an Echolance would use are photon-formed, positioned, and maneuvered vortices. Such vortices can be created because beams of light can manipulate particles in a vacuum. More specifically, lasers and microwaves would levitate and maneuver vortices of colloidal (smoke-like) particles at vast distances in front of each lance, thereby moving interstellar particles out of its path and forming some of the hydrogens into a shield. Closer to each Lance, picograms of negative matter would clear a path. Upon hitting a micro-meteorite or a proton, for instance, shield vortices of hydrogen ions, colloids, and negative matter would shatter and vaporize it. Both during and after such encounters, Enzmann's vortices would continue to function as effective shields.

Likewise, the use of significantly reduced amounts of inexpensive reaction mass and fuel would render the cost of these two items insignificant and the overall cost of an Echolance affordable. Despite its moderately expensive engines, the cost of constructing the first Echolance is estimated at $50 billion (1980 dollars) compared with the $200 billion for the first torch or pulse ship. The projected price of the second Echolance is reduced to $10 billion – much less than the $100 billion for the second torch ship, and the 250th Echolance may cost only $5 billion.

Construction of a $5 billion vessel to carry 5,000 to 20,000 people could be financed by nations, corporations, international organizations, religious groups, ethnic groups, or even individual paying passengers. Once influential groups perceive that such starships' construction is feasible, the necessary funds should become available. A mini fleet of three Echolances could be built and launched at a total cost resembling four Trident submarines or nuclear-powered aircraft carriers. Construction would take place on Earth, assembly in Earth orbit.

An expeditious program could probably build and launch the first Echolance fleet in ten years or less. A gradient plunge around the sun would accelerate an Echolance rapidly and inexpensively. Such a construction and launch program awaits no major technological breakthroughs. Particle beam starships could have been built in the 1950s and launched in the 1960s.

Time dilation at near-light velocities

The beam drives of Echolances would produce much higher specific impulse than do the engines of either pulse ships or torch ships, thereby generating enough propulsive force for a sufficiently long period to attain near-light velocities. Consequently, any of a vast number of star systems could be reached before crew and passengers age significantly.

According to Voigt's inescapable equation and demonstrated by Kaufmann, acceleration of an Echolance to more than 99.9 percent of light speed, which appears quite feasible, would reduce the experienced time and aging processes markedly. The equation for time dilation is: $t = t_0 \sqrt{1-v^2/c^2}$. Hence starship time at velocity v (in this case, the velocity of a starship) equals time at rest in the inertial continuum (normal time) divided by the square root of 1 minus the velocity squared, divided by the speed of light squared. The time dilation effect expressed by this equation has been experimentally verified, with muons measured in accelerators at CERN in Geneva, Switzerland.

An objective human life span could last for centuries of centuries, though the subjective lifespan would remain unchanged. An Echolance could attain velocities sufficient to reach stars in all safe volumes of the Milky way or cross its 70,000-light-year disc in not much more than five shipboard years. Thus, the time dilation effect that accompanies near-light velocity opens up to humans a scintillating spectrum of ventures far beyond nearby star systems.

The prospect of persistently traveling at velocities close to that of light to remain young supplies a motive to enlist as a starship crew member. Yet on the Earth, close friends and relatives of a space-age Dorian Gray would die or at least undergo extensive aging during the time required for him to complete a roundtrip to even the nearest stars. Would voyagers on long passages be haunted by bittersweet memories of loved ones from whom the time dilation effect had separated them irrevocably?

Consequences for humankind

The spectrum of possible consequences of starflight for the star bound and Earth dwelling members of the human species needs to be viewed against interrelated global problems. Revolutionary changes in communication, transportation, and trade have drawn the diverse nations and peoples of this heterogeneous world into frequently abrasive interaction. Socio-cultural limits to growth, which unnecessarily restrict the potential for sustainable growth, currently trigger shortages of the resources required to meet basic human needs and enhance life quality.

The switch in economic outlook from viewing the world economy as an 'expanding pie' to a strictly 'limited pie' generates heated controversies concerning the equitable redistribution of pieces of this pie. The narrowly defined self-interest of political and corporate decision making, the stockpiling of deadly weapons, and physical frontiers' exhaustion exacerbate pessimism, despair, and violence. Besides nourishing counterproductive animosity, a loss of vision and of nerve interferes with the formulation and pursuit of programs for spending humankind's time, effort, and resources in ways likely to gain the greatest return.

A crucial requirement of ameliorating global problems is the projection of new positive images of desirable, attainable futures. Such images tend to become self-fulfilling by motivating well planned and implemented programs to achieve the ends that the images depict. The prospect of starflight presents humanity with the glowing image of human exploration and migration throughout the Milky Way Galaxy, coupled with improving the quality of life on Earth. In short, starflight constitutes a constructive project capable of vastly expanding the opportunities of the human species.

Far from being quixotic aspiration prone to shattering people's expectations, starflight is feasible with present technology. The cost and time requirements are such that a fleet of Echolances could probably be built and successfully launched before the turn of the century. (This manuscript was written in the 1980s) Even if the cost, which is estimated at around $100 billion for a small fleet, would run to $200 billion, the program could be pursued simultaneously and contribute to the amelioration of problems on Earth. A soaring future offers not an either/or alternative but a both/and opportunity.

Available technology opens up to humankind a new geographical frontier: interstellar space with its plethora of star systems. This open frontier dramatically enhances the range of opportunities available to individuals, social groups, nations, corporations, international organizations, and the ongoing human species. The Echolance's near light velocity, together with the accompanying time dilation effect, would allow voyagers to reach any one of billions of stars while aging only a few years. Many of these stars probably have an Earthlike planet or at least a habitable planet with a breathable atmosphere and bearable gravity. The need to explore and settle such a new frontier can be expected to propel humans beyond the constraints of their Earthly island. Exploring star systems, we become genuine pioneers, using our intelligence to derive all that we need to thrive from landfalls.

Risks can be reduced but not eliminated during the exploration and settlement of new frontiers. Yet, the risks of starflight are more than counterbalanced by the prospective benefits that it offers. Fierce competition for possession of the pieces of a limited terrestrial 'pie' could metamorphose into the constructive pursuit of a nourishing and expanding 'pie.' The proverbial 'pie in the sky' is no chimera and need not be postponed till 'by and by.' The dawn of the new age of discovery and migration promises a Horn of Plenty.

Among the lustrous goals that motivate starflight are:

1) Increased likelihood of the survival and development of the human species.
2) The enhanced power of a group such as an ethnic or religious minority, an endangered nation, or an ideologically oriented movement to determine its own destiny by freeing itself from the shackles that have restricted its

development on Earth and by freeing it to become what it chooses to be.
3) The enormous wealth available from virtually unlimited resources combined with technological skills.
4) Adventure, excitement, and the thrill of beginning afresh and working to create utopian societies.
5) Greatly expanded knowledge and opportunities for ongoing scientific research.
6) Life-extension for crews aboard starships could be made luxurious with profits derived from transporting passengers and distributing cargo.

By remaining confined to the Earth, the human species allows several possible catastrophes to endanger its development, wellbeing, and in some cases, even its existence. These hazards include:

1) Collision with any Apollo asteroids that cross the Earth's orbit or with a large meteorite.
2) Radiation from a relatively nearby supernova.
3) Encounter with a galactic dust cloud that could first shadow the Sun, thus freezing the Earth in ice, and then stoke the Sun with fresh fuel by falling into it, thereby roasting the Earth.
4) The arrival of the expanding radioactive ring from the Milky Way's core in 30,000 to 70,000 years.
5) The Sun's becoming a variable star.
6) The possibility that the heart of the Milky Way may become an active quasar.
7) Large scale nuclear war, especially involving the use of a cobalt 'doomsday machine' that would poison much of the land, or a neutron bomb that could exterminate all land creatures by releasing fifty tons of neutrons into the air.

Such hazards must not be dismissed idly. Evidence indicates that the rapid extinction of dinosaurs was precipitated by a meteorite impact. Moreover, the Earth normally experiences ice ages every 250 million years, the same period of its rotation around the galaxy.

Starflight would help ensure the survival of the human species by transporting its members to many distant locations. Still, it would also create a dynamic type of environment, forcing wider boundaries of human existence. In a static environment, parameters typically continue to narrow until a relatively minor change can destroy entire species. Conversely, the expansion of parameters promotes adaptability to a wide range of environmental conditions and encourages diversification and survival. To colonize interstellar space, humans would fill new ecological niches and hence become more adaptable. In some instances, humans might eventually evolve into advanced intelligent beings who, in Herman Oberth's words, would be 'of mankind, but not men.'

The pursuit of the starflight option should lessen the chances of nuclear war. A major cause of greed and conflict is the desire to procure for oneself or favored group larger portions of what is perceived as a limited economic pie. Starflight offers abundant resources and opportunities to develop in ways deemed suitable to divergent groups. Since starflight can open up the galaxy as a beckoning cornucopia, why risk annihilating one's nation and thwarting the development of humankind by triggering a nuclear war to secure a larger share of contested resources or power over other nations? Why should ideologues blow up civilization to enforce an ideology when they can migrate to star systems much more amenable to implementing the ideology and perhaps even to the creation of a near-utopia?

Conclusion in the next issue of ENDEAVOR

The Theory of Relevance
Dr. and Joanna Enzmann

Energy is synonymous with information. The only known ways of transmitting information from location to location are via fields, photons, neutrinos, ponderable masses, waves, haptic loops, and chains. All known structures of the universe are configured of information carriers, all of which, if there are no singular sources or sinks, act concurrently as transmitters (energy or information sources) and receivers (energy or information sinks). All transmission of information from location to location or within locations is considered to be through channels, and all channels without exception have unrecoverable noise losses associated with them.

Suppose the universe is static, and there are no singularities. In that case, all configurations evolve through stages, including conception, birth, youth, maturity, old age, death, disassociation, and transfiguration, which implies configuration of the entity's information carriers (debris) into other entities. Life history may include one or more rejuvenations and/or multiplications.

Both evolution and 'spiral-like' (rather than cyclic) repetition of events in a location are foreordained by the nature of information carriers. The extent in space and time of each stage in an entity's world line is a function of external information (loosely: supply of matter, energy, information, etc.) and internal information (loosely: supply of matter, energy, error-correcting, and process codes, etc.)

It is postulated that all properties of space and time are a function of net balances between information carriers and nothing else. It is suggested that the universe is static; however, most of the formulations arrived at applying to both expanding and imploding universes.

Newtonian Physics, Lorentzian Electrodynamics in the Context of Order Theory, Information and Communications Theory as outlined above, together with the wealth of physical and cosmological observations gained over the past half-century, has led to some fundamental physical principles.

Keplerian, Newtonian, and Lorentzian kinetics theories are descriptions of highly localized neighborhoods rather than the cosmos. The theory of relevance applies to the cosmos if one looks to another local. The relative annual income between an American engineer and an oil magnate may be four orders of magnitude and irrelevant. In contrast, a five percent difference at the next desk is both relative and relevant.

All information (energy) transport is characterized by being noisy and ultimately being quantized. All observers (sinks) and all observed entities (sources) are ultimately configured of quanta.

All observations are, therefore, ultimately limited by observer ability. As ultimate observers can be defined by the application of E-number and the Time and Information existence theorems, it can be shown that there is an ultimate truncation point for any transcendental number. Manipulation of the number by an observer can have no physical meaning outside of storage of an energy series when the division of a ration like π may be carried as marks on paper, bits in a computer, etc. The divided-out number may be mapped by an observer to abstract information from his environment. In all cases, the transcendental $\sqrt{2}$ or π must be truncated by an E-number adapted to observers ranging from the simplest to the ultimate.

The above suggests a direction of research that has not been followed in the so-called hypergeometries developed from Gauss's work by Lobachevski, Riemann, and Helmholtz and does not seem to have been followed by Einstein in deriving the Equivalence Postulate from weighted Riemannian curvatures. The qualification is that geometry cannot be divorced from the observer. Geometric abstraction involves sources (transmitters), channels, and receivers (sinks). The geometry per se involves a number. The number may be equated to information, and information may be measured in Joules per bit. Indeed, there is no information that cannot be measured in these dimensional parameters. All geometric mapping of observables must ultimately be subject to the 'laws' (descriptions) of thermodynamics, even as information is.

From the ice age to the space age

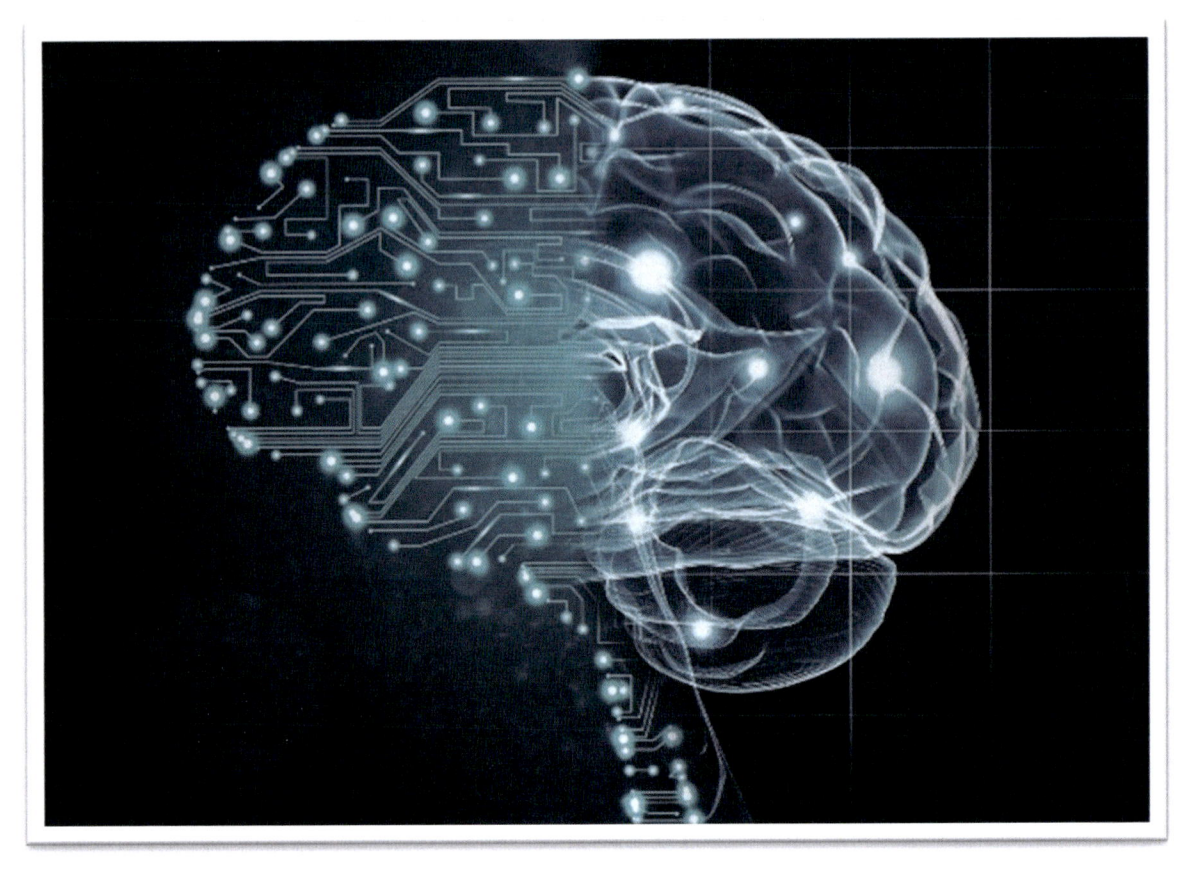

Intelligence: Educating the Gifted
Dr. Robert Duncan-Enzmann, 1949

"*On Education: With an emphasis on the careful education of the gifted, for he or she is the one that needs and can best utilize knowledge for the good of all.*" – Doc E.

One might ask just what intelligence is. What is this state which is assumed to be so important? Is it real, tangible, and measurable?

Intelligence is indeed real. Intelligence is tangible, measurable, and in the case of mankind, very important.

We might define intelligence as the ability to see the relationships between cause and effect, to see a situation in reality through an examination of its parameters, much as one theorizes on space lattices of crystalline substances by an examination of X-ray diffraction pictures or parameters, or to be able to project information and experience gained in the past to the future, to better deal with that future.

Intelligence is indeed measurable, as may be seen in Lewis M. Terman's books: *The Measurement of Intelligence, Gifted Child Grows Up, Intelligence of School Children,* and *Mental and Physical Traits of a Thousand Gifted Children*. We might think of intelligence as the function of an entity that works better in some than in others because of differences in the construction of the entities – the differences being primarily due to heredity and somewhat due to the environment.

When children are taught to read, they are usually first taught to read letters, then words, phrases, and paragraphs. Reading should be a rapid process; it should be something that is much faster than any human voice could function. Persons should be taught not to read for letters and words but solely for ideas.

One of the most beautiful devices ever thought of for teaching is now in my possession. It is a tachistoscope; it will project words, numbers, sentences, etc., on a screen for periods that will permit only one eye fixation.

Long practice with this apparatus teaches one to read a great amount very quickly and use more than just the eye's yellow spot. I also believe that it teaches one to use his memory much better and tends to develop the photographic memory. A child with a little practice, say six weeks, can easily remember 36 digits when flashed on a screen for only $1/100^{th}$ of a second, which permits only one eye fixation.

Repetition is the mother of learning, as the old Latins used to say. Indeed this is true, and things learned by rote are, when coupled with comprehension, indispensable to quick thinking.

Strangely, repetition by rote is not the only repetition that is helpful to learning. Things repeated over and over when one is sleeping is also useful and will be learned quickly when the sleeper is awake.

For teaching good music and the laws of harmony, or perhaps the multiplication tables and mathematical formulas, or real knowledge of a foreign language, a magnetic recorder placed in her room and run for several hours each night would be most valuable.

In this modern age of enlightenment, the idea of learning by rote is abhorred by many. They prefer to teach one to figure things out. Nothing could be more false. One does not increase the ability to figure things out. The I.Q. is a native capacity. This largely unchangeable I.Q. may only be brought in contact with methods that will aid and equip with systems of knowledge that will act as filing cabinets to work within stock situations.

For this reason, I recommend the teaching by rote of all the multiplication tables up to twenty times twenty. Few persons know it, and it is something needed.

The I.Q. is native. Show it generally helpful methods. Equip it with useful stock answers.

In educating the gifted, those born with a naturally higher I.Q. and therefore ability to think better than the average man – just as racehorses can run faster than truck horses – no expenses should be spared. These persons should have the best that their civilization can provide for their education – in lighting, food, desks, clothes, books, maps, etc. It is justice that these should receive the best because it is to them and not to the masses that humanity owes all its progress. To be sure, in prehistoric times and semi historic times, this progress was due to these intelligent creatures who killed the less cunning or survived disasters where the others perished through lack of cunning. This is not true now, the cunning are not allowed to kill the less cunning, and neither is nature because it would cause chaos in our mechanized world. The less brainy are needed for laborers, clerks, machinists, technicians, etc., in the intelligence pyramid that runs our civilization. The men with high I.Q.s have built a machine that the others can run; however, this machine is not perfect. It can be improved, and only the intelligent can do this work, which will better the lot of all men.

With the facts above in mind, we can see why it is valuable to all of mankind to educate the gifted before all others, so we may all benefit from their superior powers. Certainly, everyone should be educated as well as he can be, but everyone has a limit. Some persons and they are very few, have limits that cannot be measured, but most of the world's people have a very low mentality and can only absorb a very little bit of the available knowledge. It is unfair to the world to spend great sums on the defectives and neglect the superior because they can shift for themselves or because of some skewed social philosophy. This is unfair to all humanity. People should always be treated in order of their actual importance and not according to their relative weaknesses.

Teaching Language

Words change in meaning as time goes on. Dictionaries are made by using a word in ten or twelve typical sentences and deriving a brief meaning from considering them all. There are one thousand words in common use; the average man recognizes five thousand words. The very intelligent man recognizes fifty to one hundred thousand words, and there are a little over a million words in the English language.

When children have finished learning to read and are reading to learn, they begin to use dictionaries. We have some very useful and well-made dictionaries, but they are not yet all that could be desired. For one thing, the dictionaries in a schoolroom should be graded, a simple dictionary, and an intermediate dictionary, and in addition to those two a large dictionary that defines the word. The extensive dictionary should give the history of the word, its roots, Indo-European or other, the language branch it followed, such as Teutonic, Latin, or Slavic, the routes and changes it underwent, including change in its sound and the laws that it followed. All comparisons should be given in the original alphabets with translations. The book should be illustrated.

The first two dictionaries would be relatively cheap. The third would be very expensive if, indeed, it even exists. It would be worth the price, even if it were infrequently used for the reading of this work, it would be entertaining and very educational. By being exposed to such work for many years before studying philology, a smattering of the science would already be learned.

Recognition of Deadwood and Excessive Emphasis on Unimportant Collateral

It would be very healthy for the modern educators if, once in a while, they would read a history of the development of mathematics, physics, medicine, etc., with mathematics as probably the best to observe. Along with a reading of the history of the development of mathematics, educators should study the history of the teaching of these subjects. I mean precisely that they should study the old textbooks and compare them with those of today.

Such a study would lead to a revelation: Mathematics has progressed somewhat since the day of the Greeks; the methods of teaching mathematics have also changed somewhat. Instead of requiring a lifetime to study and comprehend geometry, Euclid's geometry is finished in the second year of high school. The student is released to study such things as calculus and the new algebra in college. We learn more in a lifetime!

Now, unless we concede that we have reached the ultimate in mathematics or that the human being will have a life of several centuries in a few more decades, we will have to simplify our teaching methods. The deadwood and useless proofs will have to be rejected or simplified. With a few simplifications, the calculus could be finished in high school, releasing the student for other studies.

Personal vanity plays a great part in the publishing of textbooks. Instead of being satisfied with improving the old textbooks, men are forced by their vanity to write more and more books. Undoubtedly this is valuable to science, but there should also be a body of men that look at a textbook and continually improve it through the generations. They should carefully improve the text and illustrations, simplify the proofs adding detail where the books are hazy, slowly improving the work – always and exclusively with an eye to the student's psychology. Every effort must be toward an appealing text and a book that gives the truth – symbols for nature's behavior.

Another factor that writers should keep in mind is their pupils' vanity – vanity being in men and mankind. This also will make the textbooks more pleasing.

On Writing Simple Textbooks

Hundreds of thousands of books are written every year. How many will survive? Perhaps one in one hundred years; this is true for both literary and scientific texts. However, a scientific book fragment may survive for many thousands of years because of its fundamentalness and excellent symbolic representation of nature – the real process.

The literary books survive because of their accurate depiction of man and his ways. Perhaps twenty books per thousand years will endure. The scientific books or methods, such as the Arabic numbers, survive because of their simplicity. Nature is simple, and humans try to make simple symbols for it. As soon as a valid and new symbology system that is simpler than the old appears, the ancient and complicated symbology – epicycles – begins to fade from the scene.

Great men show extreme simplicity in their thinking. It gives them an advantage. Simplicity is something that should be followed in textbooks. Everything should be explained in the sort of language that a little child could understand. The text should be something that can be read with ease and pleasure. No steps should be left out; the whole process in every case should be given.

Another factor should be considered along with simplicity, and that is constant reference to fundamentals. The knowledge of mankind is not very extensive, and it is self-flattery for a teacher to think that he is so far along the road that it is impossible to refer back constantly. He should do this. In all branches of mathematics, the very fundamentals should be reviewed from the bottom up when presenting a new problem, such as giving a lesson in tensors and building up briefly from the Pythagorean theorem.

Doc E, in His Own Words
1946 (22yrs of age)

I must not be too modest, so let's be frank. I am a genius, and a genius second to none.

Someone asked me: "When do you expect that you will receive the Nobel Prize for your monumental contributions to world literature?"

"The Nobel Prize," I snapped, "these persons don't know from nothing. Of course, I will win it, but what would a trifle like that mean to me?"

But let me say more about something interesting, that is myself. There is no reason to hide what everybody knows. I am a genius.

What have other literary giants done? You may name Shakespeare, Goethe, Anderson, Dante, Tasso – or even L'I Abner, what have they done? They have written words and had their words make sense. Anyone can write words that make sense. Chinese words make Chinese sense. Norwegian words make Norwegian sense. You see how it is. Almost any man in a million can be a great writer.

What have I done to raise my stature head and shoulders over my contemporaries?

I have created a new style – a new literary discipline, a new concept in the use of orality. Abstract writing. Yeh. That's it.

Words are chosen for their beauty, for their symmetry, for their richness in thought quality and tradition. Sordid words even become noble words through the pure spirit I breathe on them. Great works, my pearls of euphony.

There are some poor misguided fools, the pitiful idiots, halfwits, who have presumed to criticize me and hint that I have attempted to ape the modernistic painters – the painters! Phew! The painters.

As it is, some of them have a grain of sense, just a tiny grain, of course. They instinctively attempted a little pure thought, abstract thought. They attempt it in line and color, and in patches, they smear and befoul paper, canvas, wood, plaster, and sidewalks. Pencils they used, and brushes, styli, crayons, and dirty fingers – that was method! Technique, they called it. What a pitch!

Now I have come. I took some of their ideas. Did I plagiarize them? I did not. Do you plagiarize the pigs when you make silk purses out of their ears? Well, my words are silk purses full of rubies and gold.

They had common gangue; I have the gems. They had the entrails; I have a string of sausages. They have wood pulp; I have fine carvings. They have silk work squash, but I have my lady's silk hose.

Robert Duncan Enzmann on Skis in Shorts with Chessboard, Upsala, Sweden, Enzmann Archive Image

Enzmann Conference Articles

Forward, Conference One
First published in Planetology and Space Mission Planning, first conference, published by the New York Academy of Sciences, 1966.

I. B. Laskowitz
Chairman of Engineering, NYAS, and Laskowitz Helicopter Company, N.Y.

It gave me special pleasure, on behalf of the New York Academy of Sciences and as the Chairman of its Division of Engineering, to welcome the participants to the Academy's Conference on Planetology and Space Mission Planning. The appreciation of the Academy should be extended to Robert D. Enzmann, Chairman and organizer of the conference, and to all the firms, agencies, and individuals who cooperated so splendidly in making the conference possible.

Earlier in 1965, the New York Academy of Sciences sponsored a four-day conference on Civilian and Military Uses of Aerospace. It aimed to bring into focus the problems and solutions of a variety of technical developments, ideas, concepts, and techniques and embraced to a greater or lesser degree all the disciplines represented by the institutions five sections and eleven divisions. Subsequently, the United States achieved several historic firsts in space, namely Unmanned Ranger 9's 5,814 lunar photographs before impact, unmanned Mariner 4's four-day orbital flight with a maneuvering walk in space by one man with the aid of a hand-held oxygen gun, and two-man Gemini 5's eight-day orbital flight which represented the period necessary to send a mission to the Moon allow a brief stay on the lunar surface and return to Earth.

The failure of the two-man Gemini 6 mission, October 25, 1965, because of the failure of the Atlas-Agena rocket to go into orbit, made it impossible to perform the planned rendezvous and docking experiment and caused Gemini 6 to be rescheduled as a joint mission with the two-man Gemini 7 fourteen-day orbital flight. Gemini 6 made man's first rendezvous in space by maneuvering within ten feet of Gemini 7 and splashed down after twenty-six hours aloft, while Gemini 7 followed on December 18, 1965, after completion of a record fourteen-day flight. Gemini 8's historic docking with an orbiting Agena rocket on March 16, 1966, nearly met catastrophe but for the skill and courage of the two astronauts. Failure again of an Agena rocket to orbit on May 17, 1966, canceled the launching of the Gemini 9 for about three weeks. When Gemini 9 was launched on June 3, 1966, it made a successful rendezvous with an orbiting Adapter. Docking was prevented by the failure of a protective shroud on the Adapter to blow clear. Extra-vehicular activity by one of the astronauts was curtailed by fogging of his visor. Gemini 9 made a close splashdown on June 6, 1966, permitting for the first time, television coverage by cameras on the aircraft carrier *Wasp*.

Surveyor 1's dramatic success represents a near-perfect mission, from the landing on the moon June 1, 1966, through its 63-hour journey of 248,000 miles to the soft landing on the Moon, June 1, 1966, and the sending of thousands of high-quality pictures back to Earth.

After repeated failures, the Soviet Union's massive program has also scored some historic space firsts, namely: hard landing onto the surface of the planet Venus and the putting of Luna 10 artificial satellite into orbit around the Moon, April 3, 1966.

The problems that must be dealt with and surmounted in the exploration of space are still great and there is no room for complacency. What constitutes a properly balanced allocation of our country's resources into the exploration of space has been widely discussed, and it is hoped that the publication of this monograph resulting from the Conference on Planetology and Space Mission Planning will answer many of the problems raised and will serve as a reference book for scientists and engineers.

Preface, Conference One
Dr. Robert Duncan-Enzmann

The USA, USSR, and a European consortium are engaged in space exploration. Plans are to expand unmanned and manned efforts through and eventually beyond the Solar System. Equipment nearing operational status will make possible off-ecliptic, solar, cometary, asteroid, Jupiter, and Saturn probes, and manned lunar landings. Such explorations consume man-years, natural resources, and power and force economic adjustments that approach and will likely exceed the expenditures needed to conduct the First and Second World Wars.

The currently voiced objective of all space exploration programs is to gain information while protecting alien and terrestrial environments. None want a figurative space crushed through 'Dead Sea Scrolls' in a rush to 'get the gold,' as would result if an alien biosphere were destroyed or grossly altered technology in the form of propulsion, instrumentation, communications, educations, and so forth is to be spent to gain a commensurate information return. This may be done by considering the environments to be explored and the single most important aspect of technology needed to do this – propulsion.

The papers in this symposium have not been devoted to examining all these questions exhaustively; they have been selected and arranged in a fashion that it is hoped will inspire and guide the space community toward general approaches to Planetology and Space Mission Planning. Authors from the following groups are represented, with the hope that the approaches will be through finance, education, industry, and government.

Some Implications of Extrasolar Intelligence
Frederick Ordway, III

This article's references can be found in the original publication, in Planetology and Space Mission Planning, New York Academy of Sciences, 1966.

In this monograph on planetology and space mission planning, we are concerned primarily with (1) the environment of our home solar system, (2) signatures as we can measure and understand them at our present level of scientific advancement, (3) the technology now available or predictably available within the relatively near future that will enable us to undertake planetary exploration operations, and (4) the major aspects of current and near-future mission planning, considered in the light of what our science and technology offer and the extent of the physical frontiers to be crossed.

As mankind probes ever deeper into the mysteries of interplanetary space and the worlds that orbit through it, inevitably, the question is asked: "Do solar systems similar to ours exist elsewhere in the Milky Way galaxy and the billions of galaxies beyond?" If they do exist, and if life has evolved on individual planets offering suitable conditions, we can surmise that the same problems and unknowns facing our civilization have been met and solved by other civilizations across the entire universe.

Ever-increasing knowledge of our solar system has revealed that only the Earth offers environmental conditions leading to the appearance and persistence of intelligent beings. While a very primitive plant life may have evolved on Mars, the possibility that it has given rise to complex animal forms is so vanishingly remote as scarcely to be worthy of consideration. Conditions of the remaining bodies that revolved around the Sun are even less attractive than on Mars, yielding the inescapable conclusion that in our particular solar system, only one world – the Earth – harbors intelligence.

During the last five years, much attention has been directed towards the proposition that even if we are alone in the solar system, ours may be only one of many civilizations in the Milky Way galaxy – and countless billions of other galaxies in the observable universe. If experimental evidence and scientific reasoning can demonstrate the probability of what we fans call extrasolar intelligence, we may be advised to reexamine selected aspects of our long-range mission planning and planetology programs, taking into account the fact that our civilization could benefit in a very real way from knowledge gained from sources in the exterior universe.

The subject of extrasolar intelligence has been debated in scientific conferences and the pages of journals and books appearing in many countries. Hundreds of articles covering material indirectly applicable to extrasolar life have appeared since 1960.

Many compelling reasons are presented in this literature leading to the conviction that extrasolar societies are prevalent in the Milky Way and other galaxies, the most important of which are briefly summarized below.

Man has become increasingly aware of the fact that uniqueness is not a characteristic of nature. Throughout the observable universe, galaxies similar to the Milky Way extend apparently without end. Thus, on the largest galactic scale, we detect nothing special, nothing inherently different. As for our Sun, it is an average G-type star much like uncountable others – like about 15 percent of all stars in our galaxy. It contains the same elements as stars nearby and as well those hundreds of thousands of light-years away; it produces energy and light by the same mechanism as myriads of others.

Until relatively recently, the Sun was believed by many to be unique in having a family of planets. Our second point, then, is to note the significant change in scientific opinion, namely, that extrasolar planetary systems are far from rare, and indeed, may be quite common. Theoretically, planetary systems are suspected to be a normal corollary to the stellar formation, and observationally, astrometric studies have demonstrated that several nearby stars are accompanied by unseen, planetary-size companions. Recognizing that the nine planets of our solar system would be invisible to the best telescopes we have on Earth were they located on a planet circling the nearest star, the fact that we have not yet actually seen extrasolar planets does not prevent us from believing they number in the billions throughout the Milky Way.

From the above, we are persuaded that solar systems are neither unique nor rare. We have, as yet, no way of knowing how many planets are present in the average solar system, nor how they may be distributed as to size, mass, inclination, spin rate, etc., in terms of distance from their primary; but we can consider the various constraints on hypothetical habitable versus uninhabitable planets. Unfortunately, since we can only observe the solar system – our own – it is still impossible to predict whether or not the average solar system will have at least one habitable planet, but logic and statistics do not suggest such planets to be rare.

If environmental conditions on a given planet are conducive to the origin and evolution of living organisms, the problem is now to determine if life will spring forth inevitably, given sufficient time. Gain, and an affirmative answer is indicated. Laboratory tests have shown that organic compounds can be produced under conditions believed to have existed on Earth billions of years ago when life first took hold. Biological and biochemical evidence, together with reasoned speculation, compel us to believe that, under suitable conditions and given enough time, life is an inevitable result on a habitable planet. The opinion is divided as to whether or not higher intelligence will occur. And if it does, some suspect that it will not, on average, endure long periods. However, others feel it will persist over astronomically long periods and will develop inconceivable degrees of progress.

If a biological society is inevitable or virtually inevitable on a suitable planet, the question now arises as to what heights it may attain. The capabilities of extrasolar intelligence have been pondered by some authors, and a few have suggested that it may not be biological but rather mechanical. The proposition of artificial mechanical automata evolving from biological species is startling but by no means illogical. However, regardless of the nature of extrasolar intelligence (biological, biological-artificial, or totally artificial), we are immediately intrigued by the thought of determining once and for all its existence.

Several investigators have demonstrated rather conclusively that no empirical evidence of extrasolar intelligence is yet available to us, but this does not imply it is not discoverable. An entire chapter in *Intelligence in the Universe* is devoted to the subject In it, seven principal types of evidence that might be forthcoming are examined:

(1) evidence-based on artifacts on earth;
(2) evidence-based on the catastrophic results of an unsuccessful landing of an extrasolar vehicle;
(3) evidence-based on intangibles (mythology, language, religion);
(4) evidence-based on alleged flying objects;
(5) evidence-based on the detection of radio or other signals; (6) evidence-based on

discoveries that may be made on extraterrestrial worlds in the solar system[1]; and

(7) evidence-based on inference – for example, explanations of stellar phenomena as being artificially induced.

The most probable way of demonstrating the existence of extrasolar intelligence is by electromagnetic or possible optical, e.g. laser communications. A short attempt to search for and listen to signals has already been made, but no sustained effort has yet been carried out. Extrasolar signals may be aimed specifically at us, or they may interconnect advanced societies scattered throughout the Milky Way. They may even be aimed at manned or unmanned ships cruising the reaches of interstellar space. Whatever their purpose, however unusual their nature, these signals may be detectable by equipment now in existence on Earth, or equipment to be developed during the coming years and placed in operation in orbit on the Moon, particularly on the far side.

Many scientists have speculated on both the information content of interstellar messages and their 'language'. Depending on the purpose of the transmission and the technological level of the sending society, humanity may or may not be capable of deciphering its content.

My purpose in preparing this short paper has been to show, in as brief a manner as possible, why one can believe in the prevalence of extrasolar societies, to demonstrate the plausibility of the proposition that they are found through the galaxy, and to suggest that it may one day be possible to communicate at east with those nearest to us.

To my knowledge, there is no sustained program in existence anywhere in the world to attempt to locate artificial signals that may be interceptable by present-day radio telescopic equipment The U. S. is spending billions of dollars on its space program, and the Russians are spending an unknown quantity of rubles on theirs. Other nations, such as Britain and France are attempting to make meaningful contributions on their own and through European space organizations in concert with other nations. I firmly believe that all this money and effort is well spent and will produce incalculable long-range benefits to humanity. But there is one aspect of astronautics, taken in its broadest sense, that we are completely neglecting: concrete, funded experimental research of the detection of extrasolar intelligence.

The implications of extrasolar intelligence are vast. If we are convinced it exists, and if we believe it is transmitting information across the immense reaches of interstellar space (be it for its own purposes or specifically to contact emergent civilizations) we can only conclude that it may be possible to communicate with it. I strongly suspect there is an interstellar network onto which newly emergent societies sooner or later will be inevitably drawn. Societies, thousands, tens of thousands, hundreds of thousands, millions, even billions of years more advanced than ours may be reachable by electromagnetic (or perhaps optical) communications. If this is so, information gathered during the history of our galaxy may be available to us. As pointed out in *Intelligence in the Universe*, "If man is not alone, he most assuredly is not supreme. Earth-like planets billions of years older than our terrestrial orb must have nurtured civilizations faced with all the unknowns we are heir to, and most, conceivably all, of these unknowns must have been pondered, studied,

---

[1] This is to be depicted in a major MGM Cinerama film 2001 – A Space Odyssey. It is based on a novel being written by Arthur C Clark and Stanly Kubrick, and the present writer is serving as a scientific consultant. The picture is being produced and directed by Kubrick and

is being made in close cooperation with NASA and dozens of American, British, and French industrial organizations, universities, and observatories. Every possible effort is being made to ensure scientific accuracy and plausibility.

and solved by thinking beings innumerable times during the history of our galaxy. If this has happened and if the stars, like men, are born, evolve and ide, we seem compelled to conclude that knowledge is not allowed to perish but is passed on from civilization to civilization from one corner of the galaxy to the other, eon after eon."

Taking into account the enormous informational benefits that may result from an extrasolar contact, it would seem more than logical that our mission planning and our space budgets be so aligned as to make this contact possible. Only modest sums would be required to ensure meaningful progress, the results of which may be out of all proportion to the efforts made.

I recommend to mission planners and paleontologists that they take increased cognizance of the potential benefits that may enjoy as an outgrowth of research towards establishing the existence of extrasolar communications and extrasolar planets. It would appear advisable to encourage two basic types of research, namely:

(1) that which demonstrates the probability of the occurrence of extrasolar intelligence and (2) that which is concerned with actually contacting extrasolar societies.

With the first category, we can start with astronomical research. Increased efforts should be made towards detecting, principally by astrometric methods, extrasolar planets circling nearby stars. Also, further research on the origin of planetary systems and the apparent correlation between planetary systems and stellar rotation rates are called for. At the same time, research should be encouraged in biology, biochemistry, historical geology, and paleontology to further demonstrate the inevitability of the appearance of life on a suitable planet and determining the links between chemical and biological evolution. Experimental studies of the environmental constraints on life as we know it should be accelerated. Investigations of biological intelligence in man and higher animals, such as the dolphin and octopus, will provide additional insight into the basic nature of intelligence, the results of which can be correlated with stud with studies of artificial intelligence. Clearly, the services of mathematicians, computer experts, and logicians must be called upon to complement those of the biologist, biochemist, and geologist.

In the second category, concerned with attempts to detect artificial signals of extrasolar origin, research should be directed towards improving radio telescopes, antennas, radiometers, and other adjuncts of large electromagnetic communication systems, as well as pushing forward as energetically as possible on laser applications. Concurrently, the maximum feasible utilization of existing equipment for detecting artificial signals should be made, together with studies of ways and means of interpreting the information that may one day be received. And of course, transmitting (as contrasted with receiving) equipment has to be perfected: sooner or later we must acknowledge the signals we may intercept, if for no other reason than to request that information which is useful and meaningful to our civilization be subsequently transmitted by the contacted society.

Painting by Don Davis, data by Doc E

Enzmann Planetology

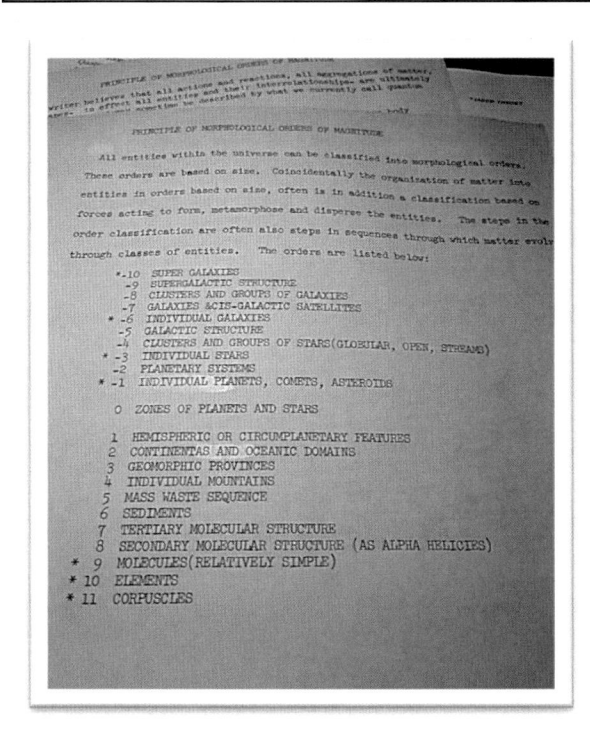

Morphological Order Theory
Robert Duncan-Enzmann

Order Theory is a systematic description of the universe extending from the smallest known entities – the fundamental particles which comprise the atoms – through a hierarchy of organized matter, to the largest currently perceived entity represented by the optical limit of the universe. The hierarchy is divided into orders, literally a cascade of 'periodic tables,' each constructed of smaller components even as atoms are comprised of electrons, protons, neutrons, etc. Molecules are comprised of atoms and so on in ever physically larger structural orders.

The sizes, shapes, and durations of things are a result of the fundamental structure of the universe forever veiled from the knowledge of Man, who may describe but never explain. We attempt to describe the essence of an electron with formulae, words, and pictures; we do not explain an electron. We describe it. We describe its essence. We are the artists. The sizes, shapes, and durations of entities of a particular order are imposed as a function of the entities' composition and energies acting thereon: external, internal, and stored.

An order and the entities of which it's comprised may be thought of as a step in a turbulent, river-like cascade. The material substance of the universe is organized into structures (orders) that endure for relatively long periods and may be compared with the horizontal steps of a cascading stream. The relatively rapid reorganization of matter to form entities of higher orders may be compared with the vertical drops of a cascading stream where the water moves from one step to another. The comparison of orders with a stream is mathematically as well as artistically valid, for the substance of the universe is caught, even as water in a stream channel, in a cascade of energy, flows from an unknown thermodynamic source to an unknown thermodynamic sink.

An example of solid matter flowing along a source-sink cascade is found all about us in the solid earth: the $^{+}1^{st}$ order stone shell of the Earth is heaved into $^{+}2^{nd}$ order continental platforms and oceanic basins by internal forces within the Earth. The continents are worn into $^{+}3^{rd}$ order belts of plains, hills, and mountains by rains driven by the Sun, even as the continents are lifted by forces within the Earth. Single $^{+}4^{th}$ order mountains are broken into $^{+}5^{th}$ order masses and boulders by frost, rains, and impacts, and these into $^{+}6^{th}$ order pebbles, $^{+}7^{th}$ order silt and so to the sea to be reconstituted again into the rocky rim of the mother continent. We, the artists, wonder if the universe itself is cyclic as a continent, or as a river rushing to the sea, to be endlessly replenished by rains drawn by the Sun from the sea and blown back over the land by the winds.

We humbly offer our pictures, words, and equations, in which we have tried to capture the essence of a beautiful cascade through time. The viewer can see that neither our words nor equations are original. We do not claim this; we wish to capture a point of view of the beautiful universe in which we live, our interpretation of the inanimate stage on which we live and play our parts.

Mensuration (measuring)
Robert Duncan-Enzmann

That which can be measured can be understood. The first man who said this was speaking of science. The meaning is even deeper.

Man must realize that his measuring instruments are not his only instruments. Man himself is the prime instrument. He measures things with himself just as all living things do. To put the thing very tersely, he is above the beasts because he can measure his environment better with his senses than they can. Take, for example, the lower organisms. A sensible amoeba could not imagine pre-cooked food. He could never measure the stove or even the food with his senses – they are too limited. He probably does not even comprehend food, as he is only drawn toward it by tropism.

To discuss the more mundane accomplishments of man's science and the extension of his senses by instruments, we may make a very general statement: That which is measured most exactly is best understood.

Mathematics is almost completely measurable as we can erect our axioms. Physics is less well known. Chemistry is not yet reduced to mechanical processes and often obscure. Biology: the present work is in biochemistry and biophysics, which have only scratched the surface.

Sociology: the study of man, it has just commenced and is under political pressure from all sorts of groups, some of which fear it, and others of which exploit a limited knowledge of man that they keep semi-secret to retain global power and extend it.

Man can be measured, and with measurement and study of man, together with the application of the results, a new era will be marshaled in. This era will mark the end of Eohomo, the dawn of man who arose through uncontrolled evolution.

A note might be added: all things cannot be measured. Even in sciences like mathematics, where we have absolute control over the environment, we cannot comprehend the environment we create on paper. The number one example is the multiple body problem. It has a simple natural solution; all equations can be written down perfectly, and yet no good solutions can be given except in special cases. We cannot understand the environment; our senses are not good enough measuring tools.

Parametrics: the Goldilocks method.
Doc E and Michelle Snyder

Three bowls, one too hot, one too cold, one just right. Three beds, too big, too small, and just right.

Parametrics are words that compare one size to another, one thing to another thing, without specific measurements.

Hot, hotter, hottest, scalding hot, impossibly hot.

Cold, colder, coldest, impossibly cold.

Small, smaller, smallest, impossibly small.

Big, bigger, biggest, impossibly big.

The same parametric usage can be applied to color – red, redder, brilliant red. One can picture red, but what intensity? One person's brilliant red might be more or less than another person's.

Another lesson to be learned about parametric measurement is that today's metrics, i.e.: 5 foot 3 inches, or 87.9 pounds, is tomorrow's parametric. A foot is twelve inches, and at one time was a very accurate measurement, yet today it is commonly used to estimate size: it's about a foot long. The more infinitesimal sizes can be, the more our past methods of measuring become parametric. Yesterday's megalithic yard became inches and tenths of inches. Today's ability to measure microns and atoms relegates yesterday's 10th of an inch to mere parametric accuracy.

Our descriptions of people are parametric – were they tall? Short? Skinny? Fat? Blonde? To be more specific, one would have to ask, "Compared to what?" Even male and female have become relative in today's culture.

Historical records are also parametric. BC and AD are very general relegations for event dating. We group events into eons, eras, millennia, years, even months, and days. All are parametric in concept. The universe moves with such astronomical precision that in order to be metric rather than parametric, complex charts must be made as to the position of all the planets and stars at the time of a particular event, such as the day you were born.

All language starts with nouns. One must have a noun before a verb, which is the movement of a noun, or even an adjective, which is the description of a noun. Parametric language is the beginning of measuring. The more technical our tools, the more specific our measurements have become.

Judgment does not escape. This article could be good, great, or wonderful, or bad, worst, or horrible. People are labeled parametrically; good people, bad people, evil people, saints, etc., are not metric measurements; they are parametric comparisons that have relative meaning to the one using them.

Have a wonderful day!

Rocks in My Head
Peyton Beals, Geology Consultant

Synthetic Pegma... *what?*

As a complete digression from this article, I was disheartened to learn of Dr. Robert Duncan-Enzmann's passing on October 19, 2020. I regret that I never had the opportunity to meet with Dr. Enzmann personally. The Enzmann Archive materials indicate that he was an individual of immense intellectual capacity and the possessor of both broad and deep concepts that span from paleolithic languages to interstellar starship design. Until there is no further reason to do so, I will continue to research the works of Dr. Enzmann and attempt to promote and distribute his knowledge.

In this installment, I will focus on one of Dr. Enzmann's professional projects that may provide an appropriate segue from my previous offering. Dr. Enzmann was involved in an effort with the title of the *Synthetic Pegmatite Project*. The definition of a pegmatite will be detailed later, but this interest in human manipulation of geologic processes could explain Dr. Enzmann's interest in the creation and production of synthetic gems (cubic zirconia) that I covered in my last article (*ENDEAVOR #6)*. Both activities indicate a desire to understand and control the natural forces necessary in the formation of rocks and minerals.

During the latter half of the 20th century, scientists and entrepreneurs sought to create and fabricate valuable materials that included emeralds, topaz, and diamonds. For whatever reason, Russian scientists appeared to lead in this activity. Russian scientists were the first to synthesize cubic zirconia, and they were also the first to create synthetic diamonds. At that time and even unto the present day, the energy required to accomplish the desired results often exceeds the eventual product's value. Apart from cubic zirconia (which is not a naturally occurring gem), the quality of synthetic gems has seldom achieved the standards established by agreed-upon mineralogical or gemology criteria.

Pegmatite is an intrusive, igneous rock formed underground, with interlocking crystals usually larger than 2.5 cm in size (1 in.). Most pegmatites are found in sheets of rock (dikes and veins) near large masses of igneous rocks known as batholiths. Batholiths are thought to result from lighter crustal material rising after being melted within subduction zones at tectonic plate boundaries. Most pegmatites are composed of quartz, feldspar, and mica, having a silica content similar to granite. Rarer intermediate composition and mafic (rich in iron and magnesium) pegmatites containing amphibole, Ca-plagioclase feldspar, pyroxene, feldspathoids, and other unusual minerals also occur. They are found in recrystallized zones and fissures associated with large, layered, ferromagnesian igneous intrusions (also batholiths).

Pegmatites are of both academic and economic interests. Academically, pegmatites represent complex geologic zones that can exhibit extraordinary properties instructive in the formation of rocks, minerals, landmasses, structural geology, and geomorphology. Economically, pegmatites are the source of valuable minerals, gems, and precious metals.

Except for diamonds, almost all other precious gems are found in pegmatitic zones. It is thought that these gems are crystallized *in situ* within the pegmatitic zone in a process that, while much more involved, is not that dissimilar from the popular science fair project of growing sugar rock crystals in our kitchens. In contrast, diamonds are expelled from regions in, or near, the earth's mantle through gaseous and explosive volcanic eruptions that manifest as the famous Kimberlite pipes. Gold and silver veins are often associated with hydrothermal deposits that reside within and at the boundaries of pegmatitic zones.

The primary issues of recreating pegmatite conditions are the original starting materials, temperature, and pressure. The problems presented are not dissimilar to those involved in the creation of synthetic diamonds except that the starting material for diamonds is always pure carbon. To artificially create a pegmatite, the starting materials (silica, aluminum, sodium, calcium, potassium, iron, etc.) must be heated to high temperatures (>600° *C* or 1,112° *F*) and pressurized at very high levels (>1,725 *bar* or 25,000 *psi*.). *(see Figure 1)*

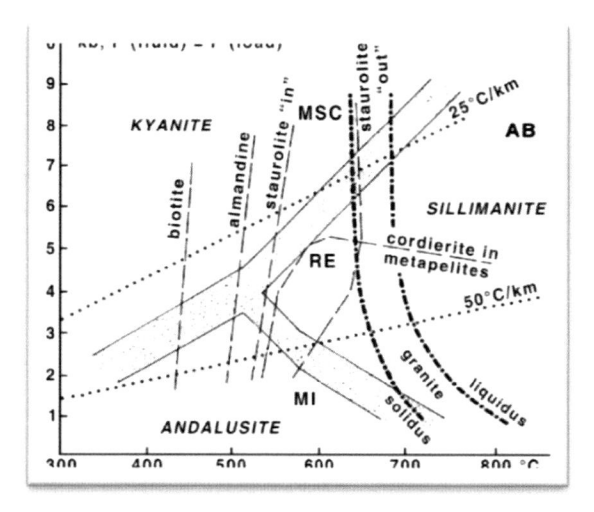

Figure 1. Pressure and temperature relationships between the four major categories of pegmatite classification, MSC-Muscovite, AB-Abyssal, RE-Rare-Element, and MI-Miarolitic (Černý, 1991).

Once the high temperature and pressure requirements are met, the resulting crystallization is determined by exothermic conditions. That is, the rate and modus of cooling determine the size of the eventual crystals. If the magma, or geologic "liquor," cools too quickly, vitrification (glass creation) occurs, and large crystals are not possible. Slower cooling results in generating large, sometimes very large crystals. However, this cooling, or curing, may not be an obvious polynomial decay determined by simple thermodynamics. It might occur as a series of stages whereby temperatures and pressures are maintained, at relatively constant conditions, for protracted periods.

The above explanation is an oversimplification of an extremely complex geologic process to which hundreds of technical papers have been devoted. Other variables in the formation of pegmatites are the water and carbon dioxide content in the magma melt. The presence of both water and carbon dioxide are believed to be necessary to promote ionic migration (to promulgate mineral crystallization) and moderate the cooling process. Modern capabilities make the temperature and pressure requirements achievable. Still, it remains challenging to simulate the exothermic conditions provided by hundreds of feet of insulating country rock and the mass of the igneous intrusion.

One of Dr. Enzmann's lifelong interests was the mineral beryl (also expressed as the gems emerald and topaz) and the industrially important beryllium metal production. For several years of Dr. Enzmann's career, he was even employed by the *Beryllium Corporation of America*. The primary source of high-grade beryl is from crystals and masses formed within pegmatites. Beryllium possesses several beneficial attributes. Relative to most other metals, it is very lightweight while maintaining structural integrity. Compared to some of the other lighter metals (aluminum and magnesium), beryllium has a higher melting point (1,287° C or 2,349° F). It is also an efficient reflector and moderator of neutrons with regard to applications involving nuclear reactions and reactors.

The above factors may contribute to Dr. Enzmann's vision of interstellar starcraft propelled by nuclear fusion-based (deuterium/tritium) engine designs. Such ships would require strong, lightweight construction materials that are reasonably immune to high temperatures and resistant to deterioration resulting from radiation exposure. For example, the *International Thermonuclear Experimental* (fusion) *Reactor* (ITER) Tokamak incorporates beryllium metal as part of the containment vessel design.

Further, beryllium oxide, a precursor to metallic beryllium, is utilized as an insulator in the construction of magnetron devices. Magnetrons are critical components of radar systems and microwave ovens and are responsible for the generation of electromagnetic energy. This may explain Dr. Enzmann's career evolution into an expert in radar technology and his eventual employment with the Raytheon Corporation. Perhaps his interest and knowledge of beryllium properties lead to this transition in his professional path. A brief conversation with Dr. Enzmann would have confirmed this supposition, but that is, unfortunately, no longer a possibility.

As best as can be discerned from the Enzmann Archive, Dr. Enzmann became involved in the *Synthetic Pegmatite Project* sometime during 1958 while employed by the Beryllium Corporation of America. However, correspondence and reports contained within the Archive suggest that the *Synthetic Pegmatite Project* was a joint effort that included the *Radio Corporation of America* (RCA), *General Electric*, *Boston University,* and perhaps other organizations.

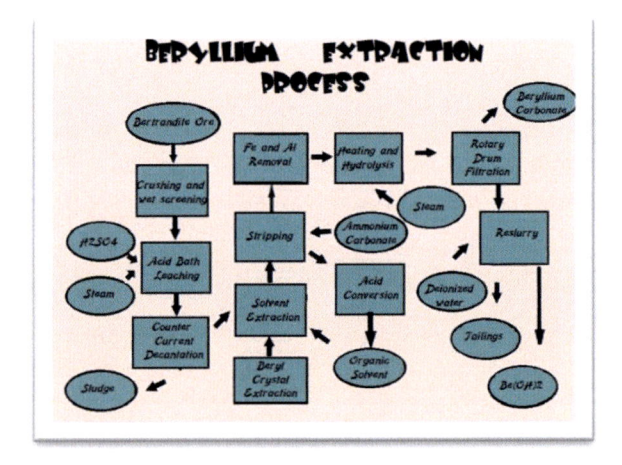

Figure 2 Beryllium extraction process

It also appears that the title of the project was a bit of a misnomer. A document in Dr. Enzmann's handwriting indicates that the project intended to extend and improve the Kawecki and Copaux process of concentrating and beneficiating beryl ores. It was less about simulating pegmatite formation and more about adopting known pegmatite conditions to aid in the purification of beryllium ore and produce pure beryllium oxide and beryllium metal.

The details of the numerous physical and chemical steps required to extract pure beryl, beryllium oxide, and beryllium metal from ores (typically, bertrandite pegmatite) are beyond the scope of this article *(see Figure 2)*.

Suffice it to say that the process is labor-intensive, tedious, expensive, and even dangerous since beryllium (dust) is toxic. Any reduction in these activities would represent significant safety and economic benefits.

Again, as best as can be determined from examining the materials in The Enzmann Archive, Dr. Enzmann's approach to the purification process was to remove several chemical modifications involving both acids and bases while also eliminating some of the necessary physical filtrations. In essence, Dr. Enzmann intended to "squeeze" impurities out of the crushed ore by employing pressures and temperatures that approached those that existed during the original creation of the bertrandite pegmatite. He then hoped to migrate any impurities away from the high purity beryllium material. A loose analogy to this would be the re-smelting of scrap metals and allowing impurities to "slag off," thus recapitulating the original metal production. It seems that Dr. Enzmann's concept was to return the beryllium ore to an earlier physical state that could then be manipulated to advantage. The ultimate result was to produce a billet of beryllium oxide with superficial mineral impurities that could be machined away.

Dr. Enzmann described this technique of hot-pressing materials. One to ten kilograms (2.2 to 22 *pounds*) of finely crushed (0.85 *microns*) ore were placed in the correct sized graphite sleeves. Graphite plungers were inserted into both ends of the sleeve to contain the material. The pressure was applied to the plungers employing a hydraulic press. The material was then heated through the graphite sleeve by employing an electrical induction coil. The samples were heated to temperatures of 1,500-2,500° C (2,730-4,530° *F*) for 4 to 10 hours under pressures of 60-100 *kg/cm²* (850-1,425 *psi*.). After this was accomplished, the samples were further cured in a kiln at temperatures of 1,000-3,000° C (1,830-5,435° *F*) to drive off (calcinate) additional impurities.

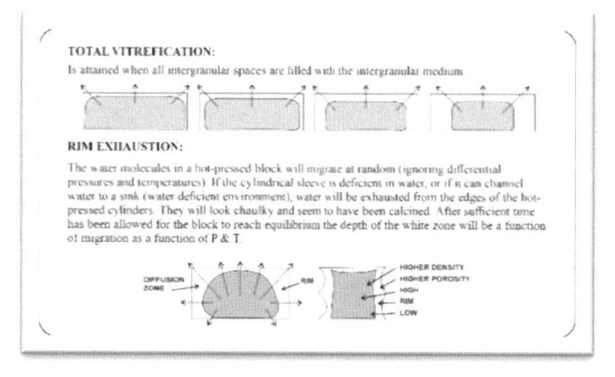

The Enzmann Archive contains both physical and photographic results from the *Synthetic Pegmatite Project*. During my inventory of the geologic materials in the Enzmann archives, several examples of "thin section" samples were uncovered. These were the size of a microscopic slide. Also found in The Archive were semicircular samples that gave the appearance of porcelain. They were mostly of an off-white, cream color that, sometimes, possessed small dark inclusions. Having already reviewed the Synthetic Pegmatite Project documents, I was able to make the relatively obvious correlation between these samples and the project. The various papers and reports associated with the project also include crystallographic and photographic evidence of the results.

Since pegmatites' primary characteristic is crystal growth, I was initially surprised by the lack of large crystals observed in the thin sections and photographs. I assumed that the experiments had "failed" due to the absence of apparent crystallization. It was only after further research that I began to realize that the goal of the project was to produce a single, monolithic crystal of beryllium oxide, of extremely high purity, with virtually no inclusions. I am, by no means, certain of this speculation. Still, it appears to be supported by the copious notes made by Dr. Enzmann regarding the formation and location of the darker ferromagnesian impurities *(see Figure 3)*.

Figure 3 Vitrification

He seemed intent on understanding how to migrate these impurities to the edges of the experimental samples. *Figure 4* is an example of what I believe to be an experimental success (on the left) and two less encouraging samples (to the right).

Figure 4 Beryllium

A perusal of modern beryllium ore processing indicates that Dr. Enzmann's approach has not been adopted entirely. However, aspects of the *Synthetic Pegmatite Project* can be found in current techniques. The process of 'hot-pressing" is utilized to create beryllium metal billets of the highest purity.

Once again, I have ventured into the mind and accomplishments of Dr. Robert Duncan-Enzmann. I am consistently impressed by his distinctive combination of a creative and adventurous approach to futuristic concepts meanwhile maintaining a high degree of attention to detail. Though the *Synthetic Pegmatite Project* may have been only marginally successful, it is another example of Dr. Enzmann's desire to expand human knowledge in the natural sciences.

Figure 5 is a close-up of the final product, which is a billet of pure beryllium.

It is possible, even probable, that the pressures described in the experiments were insufficient to obtain the desired results. It is conceivable that with more modern and competent hydraulic apparatus (diamond press), Dr. Enzmann's goals could have been achieved.

Ultimately, I feel that *The Synthetic Pegmatite Project* was both economically and academically interesting to Dr. Enzmann. Certainly, since he was employed by the Beryllium Corporation at the time, there was a compelling economic aspect to the endeavor. However, Dr. Enzmann appears to never allow an opportunity to pass in which he can accumulate additional knowledge. His notes concerning the project suggest that he also intended to prepare and submit a scientific paper on the effort. As always, I will continue to delve into The Archives in search of such items.

Notes:
1. From GreatMining.com/Mining/MetalsInfo/Beryllium
2. Kjellgren, Bengt R. F., Status of the Beryllium Industry in the United States of America (date unknown).

Enzmann Cosmology

Science and Holy Scripture
Dr. Robert Duncan-Enzmann
Dr. Rick Funk

Reverend: You'll begin with scripture?

Mathematician: Yes. Of course, with 'Let there be light'(Gen.1:3), 'from dust to dust.' (Gen.3:19) and ending with 'world without end' (Isa.45:17).

Reverend: The materialistic part of our Christian Cosmology describes life cycles of all things. There are cycles within cycles within cycles; however, as St. Thomas said: "There is s no infinite regression."

Mathematician: I venture that nothing mechanical impelled by force can happen instantly; therefore, were anything to regress infinitely, it would never happen. Even worse, in such a universe, all would be disjoint cause-without-effect, effect-without cause.

The grandest, first, and all-encompassing as far as living and other material things are concerned, is Genesis, 'Let there be light' and Genesis 'From dust to dust.' I've often said: 'Scripture is cosmology.'

Here's a State Space Diagram of 'From dust to dust.'

Reverend: State Space?

Mathematician: Things. The morphology of all things, all coherences, all energies, 'realities,' are states. Let's talk about it when we elaborate on information and knowledge when we, as best as we can, outline the nature of both state space and information space.

Genesis, "Let there be light." How surprising that practically no one notices the $2.7°$ *Kelvin* cosmic background radiation. Its energies per unit time exceed that of all stars and other luminosities in the visible universe. How simple to derive it as traffic-waves in the chaotic æther. Light, from dust.

Observable, simple, without any of the myriad of impossibilities tormenting big-bang dogma where the universe appears out of nowhere as an infinitely dense mass expanding much faster than light's velocity.

The most determined materialist, if he has at least a modicum of background in the natural sciences and physics, knows perfectly well that "no complete description can be self-consistent, and that no self-consistent description can be complete." - Gödel's Proof.

Rather than trying to pretend completeness, I do my best to be consistent, and with this emphasize to my students that underlying the material universe, there is an ultimate consistency which no human, other intelligence, with or without any combination of computers, anything in and of the substance of our material universe, can even imagine an ultimate substance.

The simple diagram above describes our material world, which is a part of a greater reality I call God, the Ultimate, Unknowable Consistency. It is a physical fact demonstrated by *Gödel's Proof* and the *Information Existence Theorem*. (Ref.: Enzmann ed. *Conf. on Planetology & Space Mission Planning*, N.Y. Academy of Sciences.)

Reverend: Infinite regression's impossible, according to *St. Thomas*.

Mathematician: Or if you protest, see *Information Existence Theorem*, *Shannon's Theorem*, and even *Classical Thermodynamics*.

It's essential that this be realized by that group of agnostic physicists who are aware that complete descriptions are self-inconsistent and that a self-consistent description is incomplete. As I like to say to mathematically inclined atheists: I'm not intelligent enough to be an atheist, which is no compliment, as it implies that I cannot transcend both *Gödel's Proof* and the *Information Existence Theorem*.

Reverend: The Christian has it on faith.

Mathematician: And is that not a compliment on his good judgment as indicated by our mathematically inclined discussion indicating faith-is-more-fundamental-than-logic?

The Ultimate Forever Unknowable Consistency is beautifully revealed in scripture and told to those who will read and study. It's a fact strengthening faith. It's fundamental to all things, creatures, and the reality of evil and good as absolutes, rather than relativism of all and any contention, circumstance, opportunity, and misfortune.

The ultimate unknowable's the stuff of enchantment that poets and romantic writers weave into wonderworks. The ultimate unknowable is also the stuff of enchantment for physicists, mathematicians, and cosmologists who realize it.

Reverend: Omar Khayyam wrote, 'There was the door to which I found no key; there was the veil through which I might not see.

Mathematician: 'A wild, weird cline that lieth sublime Out of Space Out of Time,' as Poe said.

Cosmology begins with stars; lagoons-of-light in space's dark aetheric sea. Stars- graced with vivid flares, glowing corona, veils- the zodiacal lights, and ghostly gegenschein. Stately planetary retinues, the myriad of asteroids and comets do remind one of scripture, of the Lord's 'Many Mansions."

Reverend: Genesis, all things and lives most fundamentally are unknowably conceived.

Mathematician: Life-cycles emerge from dust; from inanimate entities in the morphological orders, conception, birth, childhood, youth, maturity, old age, [*rejuvenations*] death, dissolution, reconfiguration, back to chaos. Begin with equipartitioning and traffic waves. Chaos convolves incoherence into coherences with life cycles that must always devolve back-into chaos. Back to dust, where $e = hv(1-d/D)$.

All structures, all living things, all coherences have life cycles as dynamic steady-state structures. Their extent in **xyz** space is unknown and unknowable, and their scope in time has no knowable beginning or end. There is only one force, *reaction,* which is variously expressed.

Omitting observables creates ambiguity. Nature contains no impossibilities [*singularities*] - mathematics does. Including coherences that don't exist creates singularities [*impossibilities*].

No description of the universe by an entity of the universe's substance can be complete. Complete descriptions are self-inconsistent. Consistent descriptions are incomplete.

'Let there be light' is based on a visible 2.7° *Kelvin* background cosmic radiation. All things are convoluted from light – a gentler source than a Big Bang. Big-bang cosmology begins with an infinitely dense, infinitely small universe emerging from nowhere. Big Bang cosmologists expand space, which they claim has no properties, as it's nothing, so the expansion won't really be motion. If Earth is at the center of the optical limit, then it is the center of the alleged explosion.

Invisible mass, invisible energy, invisible particles, invisible dimensions, invisible parallel universes, and other invisible 'whatevers' are necessary to correct bad mathematics. Everything in this singularity (a euphemism for a declaration that mathematically such a situation's impossible) is there to balance unworkable equations. 'from dust to dust' (Gen.3:19), and 'world without end' (Isa 45:17) describe a steady-state universe. The big bang immerses, ends in a combination of thermodynamic heat death and eternal expansion into nothingness.

On Evil
Dr. Robert Duncan-Enzmann, 1948

"*Evil is that which damages itself or acts in such a way that its future is restricted or jeopardized. Nothing else is evil.* - Doc E

There are many examples of those who do evil by damaging themselves. Individuals who are gluttonous, promiscuous, lazy, and careless are directly evil and immediately feel the consequences. Nations and other social organizations may also be directly evil. The evil usually springs from a poor organization that permits graft or a strong body that has seized the government. The damage comes through a lower standard of living. It often results in a collapse of the respective government, nation, or society, or a decrease in the relative power of the society or nation with respect to others.

Indirect evil is slower in its action; it usually jeopardizes the individual or group most often by revenge. For example, one might beat or kill his neighbor, thus gaining a temporary advantage, but his brothers and friends could inflict a much harsher punishment later. Among nations or organizations, it would be wise never to take too extreme revenge on a fallen foe or attack a smaller power unless it is inevitable that they will never recover to repay in kind.

The last evil is that of restriction. An individual usually restricts himself by laziness, wasting energy, and quarreling or fighting uselessly in his environment, losing energy that could be used to better himself. Nations limit themselves by making life hard for the intelligent and easy for the masses, be restricting their businesses and thus main their economy that much poorer, by tolerating seditious groups, and by misplaying power politics. Seditious groups seizing the government are usually responsible for insufficient power politics. An example of such politics is attacking an enemy that is too strong, creating miss-alliances, or destroying the balance of power against himself, resulting in its loss. Evil through restriction is much more serious for a nation than an individual. The individual is usually cared for by society; the society may be destroyed by federal restriction.

The above definition of evil leads to a fascinating hypothesis: No atrocity or cruelty can be called evil unless there is a possibility of revenge, wasted effort, or bad mental or national attitude caused by excesses.

Perhaps the most pervasive evil is erasing history, which affects whole cultures and humanity around the globe. 'The future of the past'; this doesn't seem to make any sense in the real world, yet it is so beautifully expressed in the *Rubaiyat* by Omar Khayyam.

> *The moving finger writes*
> *And having writ moves on*
> *Not all thy piety and wit*
> *Can change a single line*
> *Nor move a single word of it.*

An incontestable truth? Sadly not. What is writ can be unwrit. "By controlling the past, you can control the future," Dr. Enzmann often repeated. The pen is mightier than the sword applies to whether or not it is truth that is written.

It requires great power: control of the 4th estate of the media, control of schooling from kindergarten to graduate school, control of what can be published. Any who dissent, or are intelligent enough to know, are crushed or killed. Today we call it being canceled.

Contrary to the beautiful poem, past creations can indeed be changed and moved. All things are subject to the ravages of time; some things succumb to the ruthless hand of human ignorance, and some to the destructive power of hate. All information, books, statuary, even archaeology contrary to the accepted narrative are destroyed so thoroughly that it is as though they never existed. This is done to individuals, organizations, philosophies, religions, and cultures through suppression of ideas. It was done to Goddard's Grand Design.

We see statues, texts, images, flags, symbols, even people being destroyed by those seeking power even today. To change the history of a nation is to create an atmosphere of amnesia. If the past is unknown, it can be rewritten by those in power, who then contrive a new 'correct' version of the future. Call it collective evil.

History is one of the most important things to retain. We need to know our history to become better. Learning from our past facilitates a brighter future. It helps us understand who we are, where we can go, and what we are capable of achieving. We must understand what has been done so we know what can be done.

Cosmological Thoughts
Blair March

"Science without religion is crippled" is a statement made by the late Dr. Robert Enzmann; actually, the full context of the statement is that "science without religion is crippled and religion without science is blind."

I look at scientific theories, and when some of these are compared to the real world, they fail. There are many things in the religious texts we need to look at in light of the natural world and the languages in which the stories were told.

New technology brings changes to our lives. Look at what man has endured so far in his stint on this planet. Back a few thousand years ago, humanity was roaming the icy world, then the woods and forests of this planet just like any other beast. Then he developed language, tools, and other things and successfully formed a society. Some were able to do something other than feed and clothe themselves. Society became orderly from the top down.

Yet one could say we moved from one kind of a jungle to a different kind of jungle – a jungle where the predators were other humans. Today, after many evolutionary changes, we live in a technological jungle where one survives based on knowledge of technology. Unlike our long-ago history, most of us do not make a living raising food or creating textiles for clothing.

For millennia humans have used calendars to help predict the future. This process is augmented by scientific knowledge. Most of it is a simple catalog of events and recognizing patterns. By the positioning of the stars and planets and in relationship to the horizon, ancient man could determine where they were on the earth and determine storm patterns and seasons. Everything that we do or are about in our lives is done according to the season. The seasons are determined by the position of the sun and the rest of the solar systems. With geometry and trigonometry and a simple table of where the stars are, we can determine where we are on the planet. Today we have computers that tell us where we are and model potential events. Does this change the religious ideas about the past and future?

Everything that we know needs to be tested with what can be observed. For instance, we know that the duration of light guides plants and animals during the day. As the days get shorter, the animals will feel the need to hibernate, and the plants will start to go dormmate for the season. Or as the days get longer, the plants and animals move or mate. Yet, the human jungle has changed, and we are not so reliant on the natural light as we once were. We manufacture the light today that we need. How does this affect our perception of scripture?

Our clock is based on the patterns of the solar system. Clocks, unlike the solar system, have limitations on their accuracy. In the early days of the ship's chronograph, a log was kept, and how much time the ship's clock was gaining or losing per day was recorded. If it was reset in port, a new logbook was started with the same information brought forward. Although not a daily occurrence, it happened regularly.

So, what about science and religion regarding time? I often ask, in the beginning, how long was the first day? And would we be able to measure it today? The religion that we enjoy today is not a pure religion of any kind but a combination of diverse cultures coming together and morphing, civilizations such as ancient Greek, Roman, Egyptian, Norse, and other societies. Many stories are the same; the names and language were changed to bring them up to date in a new culture. Does science evolve in this way? If yes, we could ask, is it science? Or is it the popular theory of the day?

How do science and religion temper each other? Genesis tells us what happened on each of the seven days of creation. Is there science that confirms these events? According to Doc E, there is, and he had begun compiling such a manuscript. What does science say about the soul? About the cycle of life? Our ancestors observed all these ideas, and these observations resulted in stories and images depicting the complex processes of life. Religion and science should once again become balancing forces; science should strive to confirm scripture, and religion should function within the bounds of science – actual science, not scientific opinion.

Enzmann Chronology

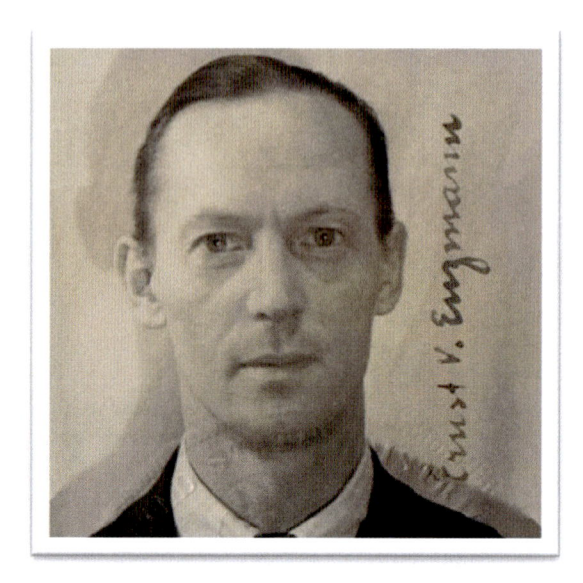

The Secret Life of Ernst Enzmann
Dr. Robert Duncan-Enzmann
Michelle Snyder

Ernst von Enzmann, my father, honored all debts, favors, and kindnesses. This is an extremely important ethic.

Father had extensive, detailed, and broad battlefield experience with all types of artillery of his time. On Kwajalein, he excellently advised and instructed the Japanese on effectively defending and destroying an attack on Kwajalein Atoll by massed battleships and naval aircraft. His work was excellent.

The Japanese spent lavishly, cutting trenches in the broad tidal seaward reef of Kwajalein. Had America attached across the mile-wide seaward reef, we should have been decimated, slaughtered, and soundly defeated in that maze of almost invisible trenches.

Father omitted to tell the Japanese, but thoroughly, with many photos at various tidal levels, showed Americans how Kwajalein could be attacked *inside* its lagoon. The American's attack on Kwajalein, called operation Flintlock (I participated in a US Navy Aircraft), was carried out just as Father had suggested in 1944. The battle, swiftly shattering Jap defenses, was decided within an hour.

Loose lips sink ships.

But 1944 is long past. In 2021 the island is so peaceful; the battleships are museums or scrap. The papers published after the attack said one US Navy battleship, but I counted four battleships with hundreds of missiles from the air. The sea bombardment was awesome (note that battleship 16-inch guns had a maximum range of 30 miles). The landings of the US Army and Marines were dominant within an hour.

Father had honored all debts, favors, and kindnesses from a past event, one that saved his life. He had just escaped Siberia, where he had been imprisoned, a story told in *Siberian Prison*, published in 2014. It was cold, and Ernst was very hungry and somewhat ill. He was hitching a ride to get away from the Japanese; they were expanding their grip on Manchuria as they prepared to invade and occupy Nanking and eventually all of China. Father staggered into the American Embassy, begging for help.

The American Ambassador bathed and fed father, put him to bed, cleaned the sleeper's clothes, and changed his Russian money for American dollars. He revived my father. It was a life-or-death intervention.

Kwajalein Atoll

Father honors debts, favors, and kindnesses. The Japs thought Austrians were Germans. Austrian by birth, father chatted in Russian, German, and English with Japanese officers about artillery. They sent him to Kwajalein Atoll to analyze naval, and land-based artillery defense and attack on Kwajalein Atoll, which the Germans had purchased from Spain with the Truk Atoll and Northern New Guinea – and the Japs seized in World War I.

And so, father repaid the Americans by spying on the Japanese military at Kwajalein Atoll, sending them photographic information to help them succeed in obtaining it as a base. Joanna and I have spent many years working there on missile defense and radar. Currently, it is governed by the Republic of the Marshal Islands.

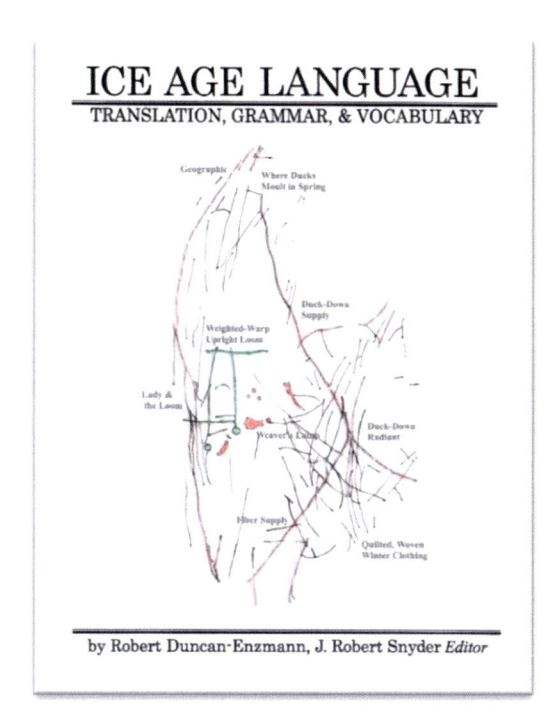

Ice Age Language; Translation, Grammar, & Vocabulary, Dr. Robert Duncan Enzmann and J. Robert Snyder, White Knight Studio, 2013

Ice Age Language
By Dr. Robert Duncan-Enzmann

This book is dedicated to the "alive" machine-learning program, ALTHEA. In it, Bob explores language patterns through tens of thousands of pre-historic centuries. He used this matrix of linguistic patterns for programming the first military learning-machine decades before.

These inscriptions tell about mothers and children, hearth and home, tools and textiles, hunting and fishing, health and medical, calendars, and contracts. One of the most essential points cryptologist Duncan-Enzmann makes is that all writing begins with sequential arrays of symbols. This observation can be derived from his analyses of Magdalenian writings. In brief, writing began with a symbol. *Major Cardinals* are *recognizable images*, which show what the story is about. In a record of how to use a horse for food, clothing, etc., a horse's outline is clearly visible. Inside the Horse Cardinal are calendrics and instructions, the when and how.

The translations of these image-stories tell us that the subject of Paleolithic writing was centered on textiles. Much, indeed most, of the Magdalenian writings concern textiles, likely all of it written by women. These translations bring us records of women making and trading textiles, and childcare. Long before we had electricity, today's versions of heating, laundry, cooking, lighting - all necessities - existed in other forms. We are not significantly different today. Most of what was written during the Bølling concerns the same things that are important today: seasons, childcare, textiles, cradles, diapers, and clothing.

Ice Age Mother & Child
This story is of the mother, by the mother, for her child. It's about who is involved, what is used for materials and tools, when those materials and tools were used, where we hunted for and gathered those materials and tools, and how. We begin with a wedding picture called *The Bride* and follow young Lorelei through her birth, her crib, her toddling years, and her late-winter expedition with the maiden, mother, and matriarch, then come around a whole generation to her engagement. The

story is told in stone and still readable today with these translations. Before civilizations can prosper, before regions can develop, before clans can flourish, before families can grow, babies are born. During a deep-cold Ice Age, ladies used looms to keep the children alive and warm. This story is timeless; the characters are us, technology today is a result of how we survived, and it's mostly written in stone by mothers - the Ice Age Mother & Child.

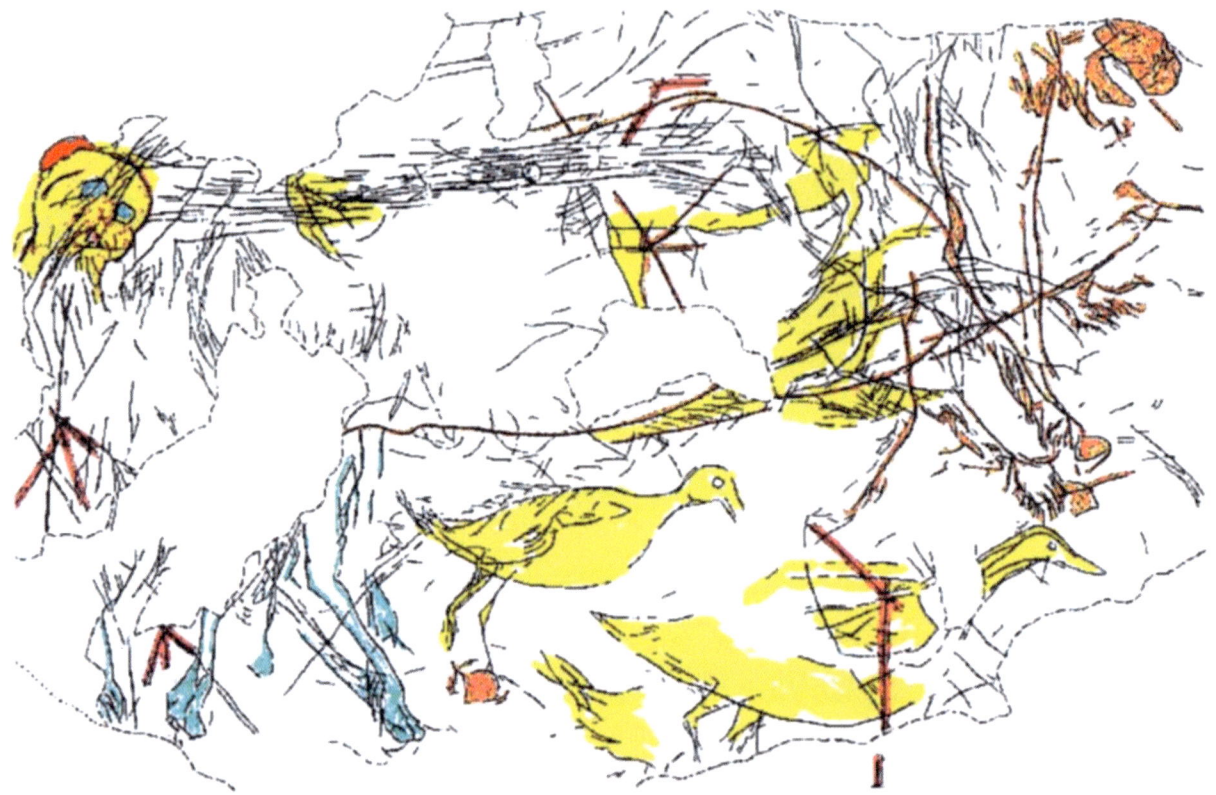

Ref.: Platte 168, by Gisela Fischer, Die Menschendarstellungen von Gönnersdorf Der Ausgrabung Von 1968. Ducks & Little Lorelei [TRANSLATION]What, when & where

The above is part of a story from 14,500 years ago about a little Rhine maiden. It includes her birth, making her crib, her quilt, her diapers, her winter outing, and her sanitary care. The story tells of her weaving by lamplight at a loom - all in the fantastic Gönnersdorf archive, written and illustrated during a Late Magdalenian, Bølling mild interval.

Lorelei sits watching ducks swim in the Rhine shallows. The narrator tells what will be done this year to keep the child clothed, fed, and comfortably warm during the coming glacial winter: 1) spring, ducks, geese; 2) summer, horses; 3) fall, antelopes; 4) winter reindeer.

In mid-summer, migrating flights of summer ducks and geese number in the billions. Flocks of winter ducks that do not migrate number hundreds of millions. Bird food - vegetable, insect, small animals, and fish - is unparalleled and bountiful during centuries when the long-grass cold prairie dominates the world's flora. It's spring, but with four solar azimuth-V's beside four secondary cardinals.

South is at the top. Bølling astronomers and map makers oriented toward the sun. Solar azimuth V's are highlighted above in red. They are linked with the following cardinals (readily recognized pictures), indicating calendrics when the animals are present and in a particular condition or activity.

The upper register is a lead-lane of symbols extending from child to horse. It details the production of adult's and child's clothing. Symbols about major cardinals are connected with link-lines. Major cardinals are associated with lead-lines.

The lead-lane of cardinals and symbols, about and between, represents the Rhine River, the shallows where ducks are caught, and the nearby ford used seasonally by migrating horses and other herds (left of the horse's head).

1st CALENDRIC-SPRING LITTLE GIRL'S RADIANT SPRING-SOLAR Az-V

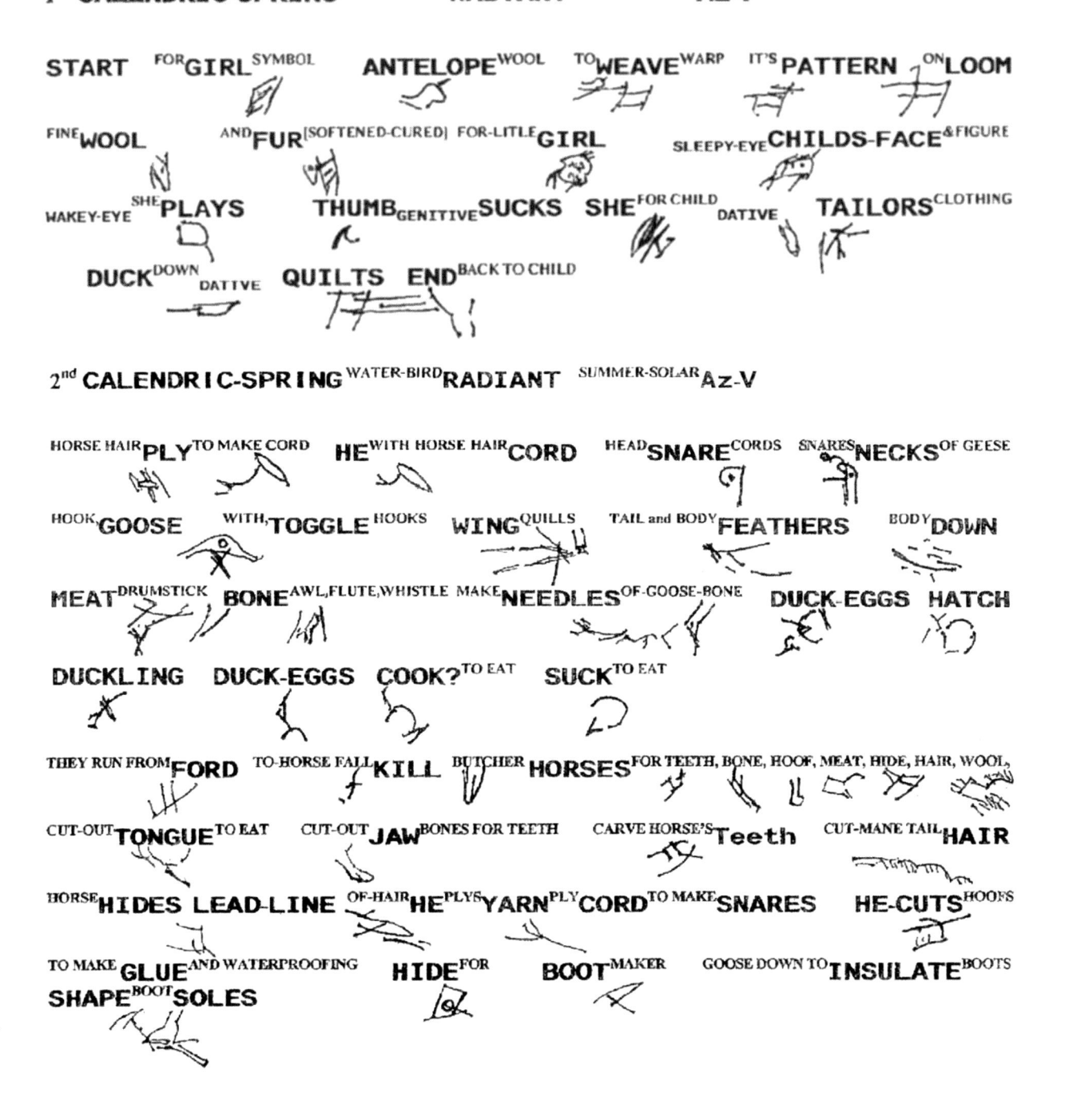

START FOR GIRL SYMBOL   ANTELOPE WOOL   TO WEAVE WARP   IT'S PATTERN ON LOOM

FINE WOOL   AND FUR [SOFTENED-CURED] FOR-LITLE GIRL   SLEEPY-EYE CHILDS-FACE & FIGURE

WAKEY-EYE SHE PLAYS   THUMB GENITIVE SUCKS SHE FOR CHILD DATIVE   TAILORS CLOTHING

DUCK DOWN DATIVE   QUILTS END BACK TO CHILD

2nd CALENDRIC-SPRING WATER-BIRD RADIANT SUMMER-SOLAR Az-V

HORSE HAIR PLY TO MAKE CORD   HE WITH HORSE HAIR CORD   HEAD SNARE CORDS   SNARES NECKS OF GEESE

HOOK GOOSE   WITH TOGGLE HOOKS   WING QUILLS   TAIL and BODY FEATHERS   BODY DOWN

MEAT DRUMSTICK   BONE AWL, FLUTE, WHISTLE MAKE NEEDLES OF-GOOSE-BONE   DUCK-EGGS HATCH

DUCKLING   DUCK-EGGS   COOK? TO EAT   SUCK TO EAT

THEY RUN FROM FORD   TO-HORSE FALL KILL   BUTCHER HORSES FOR TEETH, BONE, HOOF, MEAT, HIDE, HAIR, WOOL,

CUT-OUT TONGUE TO EAT   CUT-OUT JAW BONES FOR TEETH   CARVE HORSE'S Teeth   CUT-MANE TAIL HAIR

HORSE HIDES LEAD-LINE OF-HAIR HE PLYS YARN PLY CORD TO MAKE SNARES   HE-CUTS HOOFS

TO MAKE GLUE AND WATERPROOFING   HIDE FOR   BOOT MAKER   GOOSE DOWN TO INSULATE BOOTS
SHAPE BOOT SOLES

5th CALENDRIC-WINTER REINDEER RADIANT   WINTER-SOLAR Az-V

LOW-LEFT REINDEER   COLLECT Antlers   FAT-RICH Bones   PLUCK Wool   SCRAPE Hair

OF REINDEER SPIN   TO QUILT WITH DOWN SHE   THE WARP   SHE SETS Weaves HAIR

DEER-FOOT, DUCK-HEAD, DUCK-TAIL   ↑ up to QUILT   horse-hair   duck-down   duck-goose-down/feathers
reindeer-wool QUILT   LINK-LINES TO SEWING-RADIANT

HAUNCH-MEAT   FROM SENTINAL AT AMBUSH   KILLS HE   SKINS   RACKS AND SCRAPES

TO PREPARE FINE ANTELOPE FUR   COLLECT ANTELOPE   FLUFFY-FALL WOOL   THEY WEAVE   FOR CHILD
DOUBLE-LAYER FOR QUILTING

FOOT   SUMMER Az-V   HAUNCH Meat   SEWING 8th RADIANT   LOWER-LEFT   leg bones

[TRANSLATION] START   AT HOUSE   SHE-SITS   AT LOOM   WEAVES   SET OUTER

HORSE-HAIR   [VERTICAL] WARP   OF ANTELOPE WOOL   SET INNER [VERTICAL] WARP WEAVE

HORIZONTAL WOOF   FISH-OIL   CUT DRY FISH   RENDER GOOSE GREASE LAMP   LAMP WICK

SHE QUILTS WOVEN CLOTH   STUFF QUILTING-CHANNELS   LORELEI NAKED-BOTTOM [FROM WAIST DOWN].

WAIST DOWN]. QUILTED-TOP [COAT THIGHS UP]   BOTTOM HAS DISPOSABLE PADS   VOCATIVE-EYE
I'm being changed; I'm wet. I'm messy. It's chilly

DISPOSABLE [note: *I cannot resist saying: "BIOGRADABLE !" environmentally-fashionable and politically correct 14,000 years later. Yet, one must observe that as more and more bacterial discoveries come to light, more and more disposable things are "biodegradable."*]

HE   FISHES   SNARE DUCK[s]   SPLIT FISHES   DRY FISH   OF ANTELOPE

WOOL   TO WEAVE   FOR GIRL

In Fall, flocks arrive, and there is a Fall hunt. Winter Flocks come, then a Winter hunt. In Spring, there are eggs and a distaff for quilted Winter clothing.

The translations are transliterated into current American-English Grammar. However, symbols along lead-lanes, lead-lines, and link-lines are deliberately sequenced so that Nostratic grammar, predicted by paleo linguists as having been in use about c. 12,000 BC, is confirmed by these writings from Bølling centuries of c. 12,500 - 11,800 BC. My dictionary includes many Nostratic sequences from Gönnersdorf Platte 168.

Vocabulary
ambush [noun or verb]
antelope, antelope-wool, antler,
Az-V [Solar Azimuth V]
bill [noun - of duck] goose
bone
boot [noun]
bottom [noun - quilted winter child's pants]
child [or toddler]
cloth [woven]
coat (top winter [child's quilted] garment)
cook [verb - cook meat]
cord [noun], cut [verb]
disposable [adjectival, can be a noun - disposable moss, felt, straw to keep infant/toddler clean]
down [noun - duck or goose down]
duck, duckling, egg
END, of story
eye [noun]
face [noun - a child's face]
fall [calendric]
feather [duck or goose]
fish [noun], fish-oil
foot [noun]
ford [noun - across Rhine River]
fur [noun]
girl [noun]
glue (waterproofing) [noun]
goose, goose-grease
hair [noun]
hatch [verb - hatching of an egg]
haunch-meat [noun]
he [pronoun]
hide [noun - animal's hide]
hoof, horse, horse-hair, horses-hoof
house [noun - winter house]
insulation [noun]
kill [verb]
jaw [noun]
lamp, lamp-oil [noun]
lead-line, link-line, [syntax]

loom [noun - upright, weighted-warp loom]
Lorelei (a name taken from legend)
Meat [noun]
mouth [noun - child's mouth]
naked [adjective]
neck [noun - of goose or duck]
needle [noun]
pad [sanitary for baby, toddler]
pattern [a woven pattern]
pluck [verb - pluck duck or goose down]
ply [verb - ply thread, yarn, cord]
quill [goose or duck]
quilt [verb and noun]
quilted-top [noun - winter garments]
radiant [lead and link lines]
reindeer [noun]
render [extract oil from fish, ducks, geese, and other animals as oils, grease, tallow]
Rhine [noun - river]
River [noun]
scrape [verb - scrape hide]
set [verb - set the vertical warp of a weave]
sew [verb]
shallows [noun –along a riverway]
shape [verb]
she [pronoun]
shred [verb - horses hoof]
sits [verbal – she, the weaver, is sitting]
skin [verb or noun]
sleepy-eye [adjectival]
snare [verb or noun]
spin [verb]
spring [clendric]
START, of story
strip [verb - strip feathers and quills]
sucks [verb - child sucks her thumb, verb - suck eggs]
summer [calendric]
tailor [verb]
terrace [noun Rhine River terrace]
thumb [noun]
toddler [or child]
toggle [noun - toggle hook for catching ducks and geese]
tongue [noun]
tooth [noun]
top (quilted coat) [noun - child's coat]

vocative-eye [an alert, vaguely anxious child]
wakey-eye [adjectival]
warp [noun - vertical threads]
waterproofing (glue) [noun]
weave [verb]
web-foot [of ducks, geese]
wing [noun - duck and goose-wings]
winter [calendric]
woof [noun - horizontal threads]
wool [noun]
yarn, [noun]

During the Bølling, ducks, geese, cranes, storks, herons, terns, and other birds were vastly more numerous than anything seen since - even the North American passenger pigeons - enough to block out the sun. - RDE

The Immortal Child
Dr. Robert Duncan-Enzmann
Joanna Enzmann

Circassian Helen in The Forest World

A strikingly beautiful child plays alone in a meadow talking with imaginary friends. Ash-blonde, her pink complexion slightly tanned, the child's sapphire eyes see a wonderful world woven of colored grasses, twigs, flower chains, and imagination.

In the vast forest, sounds are soft and muffled. Light is subdued. Silent lonesome rivers wind silvery threads within it. Here and there, rocky prominences rise above the trees. Open spaces are few, far between, and small. Most are fire glades formed by lightning. Here, golden sun-warmed grasses graced with wild meadow flowers glow in the sun.

Happily, the forest child, perhaps eight years old, sings softly. What a blissful day; not hot, not cold nor hungry, neither hurt nor tired, chores done – her leather bag bulges with pinion-pine nuts. Time lingers, distilled into pure contentment in the sunshine.

She is from a hunting clan, for that's all that lives here. The child is "woodsy," which means being attuned to all that happens in the quiet forest world. The only sounds are those of nature. She listens to winds blowing about the tops of the redwoods. "Zzzzzzzaack," says a wind-flurry of hot air above the forest. "Swoonssss" valley wind from the great river, scores away, can be heard.

Smells are clean in the pure sparkling air, aroma from treetops, midway foliage, hot ground pine – the fragrance of nature surrounds her. Sometimes the dusty, impeding rain is smelled above all other scents. Other creatures – animals that are friendly, dangerous, or indifferent – come and go, warily eyeing whatever moves. Helen is no different; she remains alert. Then, she feels a far-off sound of heavy running on deeply bedded pine needles – a tremor, some crashes, then the rustling of a swift runner.

The child jumps back and wiggles into a rock cleft just as a massive spear scores rock inches away. Howling with unsuppressed emotions, hunch-thing races forward snatching up stones, and with tremendous power hurls them at the crevice, attempting to break an arm, a leg, the spine – but not to kill. No, not yet.

His face is contorted with the desire to play with his victim's body, as do wildcats with helpless, crippled animals before they kill them. Hunchman loves death play. But cats kill cleanly. He lustfully gratifies himself, then tortures little victims. Hunchman relishes messy death play. In seconds he'll cripple the child, and then fun torture will begin.

Heart pounding, the forest child, leaps like a coiled spring and frantically squirms past an angle in the cracked boulder, out of direct line from a powerfully flung shower of rocks. She shrinks back as rock splinters ricochet inward.

Phobia: it's why mammals hate spiders; in eons past, ancestors of spiders ate us. Hunchman's mind is as alien as a spider's, yet Helen is not defenseless. Now instinctively, she fights back. Goosebumps ripple over her arms, legs, and body. Her eyes widen. She breathes quickly.

Hunchman's leering face contorts. The bombardment ceases as his enraged shrieks of pain rise then fall. Extreme nausea grips him, forcing projectile vomiting. Then with agonized gurgles, Hunchman runs off into the forest racked with dry heaves. The kill has failed. He does not know about Helen's ability to vasten. Not yet.

Razor-sharp instincts and raw intelligence saved the little life. Wearing only moccasins, the child, perhaps eight years old, wiggles from betwixt the boulders and starts homeward. Urinating now-here then there such that weak breezes will loft the scent just-so, creating a false trail, combined with a curious Kirlian-like aura, save the child's life.

It's like the stench (unknowable by humans) of an adorable sand-cat kitten, a scent that violently repels all but its mother and immediate siblings. It must be remembered that canine and feline sense of smell exceeds ours by a million-fold. (Imagine a combination of skunk, burning rubber, and the unmentionable multiplied to such degree), and then there's her aura.

Autumn skies are deep indigo. Late afternoon blends imperceptibly into evening in far northern climes. Swiftly, silently, Helen speeds through the forest. She carries a leather bag of pinion-pine nuts to a withered hag, a grandniece twelvefold. Helen knows, she remembers most things, she learns quickly. Old, arthritic, vision dimmed, even at long last with swollen lower limbs, the hag's dying love is for Helen, who's somehow her child. The ancient crone will not last the coming winter.

Helen has eternal neoteny. Short are the snouts of baby animals; long and vulpine are snouts of big ones. Neoteny (babyishness, childishness) is lovable. God created love for little ones in man and beast. Even the vicious badgers have been known to adopt lost toddlers, nurture them and feed them. The forest Hag, now old and dying, loves Helen passionately. They eat broiled varmint, pinon-nuts, sunflower roots, greens with pesewa seasoning, and salt from the halite lick. Their fire glows warmly before the lean-to. White dogs lounge about them like an island of comfort in the endless forest.

The clan is gone, leaving the hag to die rather than burden their winter. Strange, Helen won't leave the decrepit husk of a female who has cared for her so long. There are weeks ahead before the snows. If each day acorns and pinon nuts are collected, they can eat during the snow moons. They can live. Or, if the hag must die, she can do so in comfort.

Swiftly, day after day, Helen, the forest child, the origin of the Dryad legends, collects acorns, nuts, and pinons. They have a wildcat kitten; they have green snakes and little barn owls perch above their burrow amidst the giant tree roots so rodents will not steal their food.

Helen is stronger. Hunchman cannot approach closer than twenty miles; besides, the clans hunt him. Yet, Helen wanders too far in her search for stores to collect for winter. Hunchman enters the camp, beats the hag, puts out her eyes, breaks the bones in her hands and feet, sets fire to the stored food, and runs off - away from the essence of the wretched girl. One day he will kill her.

Symbology

MIDAS DAUGHTER TURNED TO GOLD

King Midas Unlocked
Michelle Snyder, The Symbologist

Today, the story of King Midas and his golden touch is a tale of greed and its consequence, and in some versions, the rewards of repentance. Here is how the first part of his story is generally told today:

Midas was a king of great fortune who ruled the country of Phrygia, in Asia Minor. He had everything a king could want. He lived in luxury in a great castle and shared this life of abundance with his beautiful daughter. Even though he was wealthy, Midas thought that his greatest happiness was provided by gold. He would spend many days counting his coins.

One day, Dionysus, the god of wine and revelry, passed through the kingdom of Midas. One of his companions, a satyr named Silenus, got lost along the way. Silenus got tired and decided to take a nap in the famous rose gardens surrounding the palace of King

Midas. He was found there by the king, who recognized him instantly and invited him to spend a few days at his court, where they feasted and rested. After that, Midas took him back to Dionysus. The god of celebration, very grateful to Midas for his kindness, promised Midas to satisfy any wish of him.

Midas thought for a while, and then he said: "I hope that everything I touch becomes gold."

Dionysus warned the king to think well about his wish, but Midas was positive. Dionysus could do nothing to dissuade him and promised the king that his wish would start the following day. Everything he touched would turn to gold.

The next day, Midas woke up eager to see if his wish would come true. He extended his arm, touching a small table that immediately turned into gold. Midas jumped with happiness! He then touched a chair, the carpet, the door, his bathtub, a table, and so he kept on running in his madness all over his palace until he got exhausted and happy at the same time! He sat at the table to have breakfast and took a rose between his hands to smell its fragrance. When he touched it, the rose became gold.

"I will have to absorb the fragrance without touching the roses, I suppose," he thought in disappointment.

Without even thinking, he tried to eat a grape, but it also turned into gold! The same happened with a slice of bread and a glass of water. Suddenly, he started to sense fear. Tears filled his eyes, and at that moment, his beloved daughter entered the room. Hugging Midas, she turned into a golden statue! Despaired and fearful, he raised his arms and prayed to Dionysus to take this curse from him.

The god heard Midas and felt sorry for him. He told Midas to go to the river Pactolus and wash his hands. Midas did so: he ran to the river and was astonished to see gold flowing from his hands. The ancient Greeks said they had found gold on the banks of that river. When he turned home, everything Midas had touched had become normal again. Midas hugged his daughter and decided to share his great fortune with his people. His people led a prosperous life, and when he died, they all mourned for their beloved king.

There is more to the tale, none of it complimentary to the king. There are a few variations of details; many tellers add morals and speculation to the myth to teach virtue. This story is layered with cultural additions; turning his daughter to gold was a later palimpsest, told by Nathanial Hawthorne in 1852.

Historically there were several kings named Midas; the standard reference is King Midas from the 8th century BC, who ruled Pessinus, Phrygia. He was married to a Greek princess, daughter of Agamemnon. According to Greek and Assyrian sources, he traded with the Greeks and many other nations.

Was it Greed? Or Science!

To decode this legend, we must look at the electromotive series.

History tells us the Greeks found gold on the riverbanks, where Midas washed the curse from his hands. This may be the most significant statement for decoding this story. If you take a tin cup and place it in water downstream from a copper mine or processing plant, in a while, you will have a copper cup with the proper liquid medium. Iron turns to copper, copper to silver, and silver to gold. This process is called the electromotive series and is, in general, how batteries work. In the Smithsonian Museum, there are two silver Llamas that fell into a river where silver trace was, and they turned to silver, well-preserved to the detail. The electromotive process is symbolized by Midas turning everything to gold and washing the gold off in the river.

It is also essential to look at what was happening at the time of Midas. Turkey used young children to mine tin, as they were smaller and could go deeper into the mines. Some died and were left in the mines. If the mine collapsed or flooded, the children were left there. One mine did flood with water that held gold sediment, and a young girl about five years old was preserved as electrum: a silver & gold child. Pristine details remained: Her eyelashes were preserved to the hair, and the flowers in her basket still had their petals. The story of Midas turning his daughter into gold likely comes from records about the discovery of this young girl.

electro-chemical series

Potassium	-2.92
Calcium	-2.87
Sodium	-2.71
Magnesium	-2.37
Aluminium	-1.66
Zinc	-0.76
Iron	-0.44
Tin	-0.14
Lead	-0.13
Hydrogen	0.00
Copper	+0.34
Silver	+0.80
Mercury	+0.85
Gold	+1.68

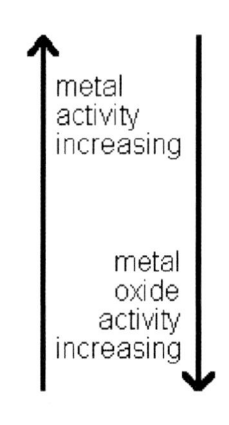

↑ metal activity increasing

↓ metal oxide activity increasing

So how did the story become so detrimental to the reputation of King Midas?

Another significant thing to know about Midas was that he invented coins for standard measurement. Well, his wife conceived the idea, and he had them made. Great! You might say. It was. Trade of precious metals and stones was big business, and it was full of thieves and cheats. The coins prevented Midas and others from being defrauded in transactions, and the swindlers hated him for it. The king's reputation became one of greed and avarice, a reputation that spread rapidly as stories made up by angry traders. In the stories about his daughter, she has been called Marigold (a name derived from Mary of the Catholic tradition). That name would not have been used then. If he had a daughter, more likely, her name would have been Elora or Lura.

Palimpsest about many legendary and mythological characters arise from similar slander. The monster Medusa was a result of stories told to tear down the power of the Vanir navigators. It is said that Athena turned her into the snake-headed creature – in theory, she did. Stories are powerful. They can create a persona that is hard to correct. Peeling away layers of information and looking at each in historical context helps determine what is valid and what is layered into history for deconstructive purposes. Midas was subject to just such storytelling.

A man preserved as copper in a Danish Bog.

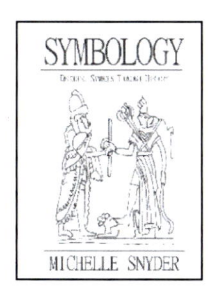

Symbologist Michelle Snyder's thesis as submitted to the University of Wales: *Decoding Symbols through History*. It could be called *Seeing History through Symbols*. Illustrated, glossary, bibliography.

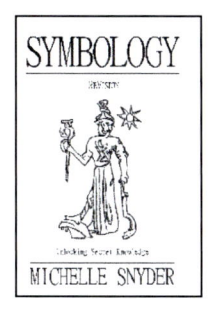

ReVision is a unique and enlightening book on symbology and symbols, based on Symbologist Michelle Snyder's work with Doc E. Illustrated. Includes a brief encyclopedia of decoded images.

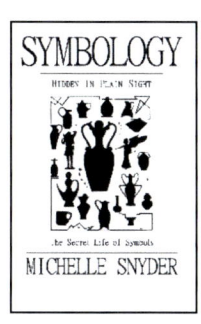

Hidden in Plain Sight is a collection of symbol systems such as heraldry, magic, the Tarot, and specific subjects like the Green Man. Symbologist Michelle Snyder takes you through the meaning and history of the symbols as decoded with Doc E. Illustrated.

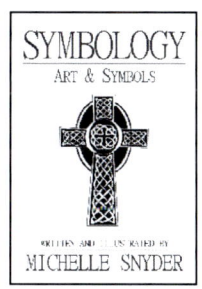

Artwork & Symbols is a book of Symbologist Michelle Snyder's artwork. It is a collection of her illustrations and symbols, with brief explanations. Her work has been shown in galleries from California to Massachusetts.

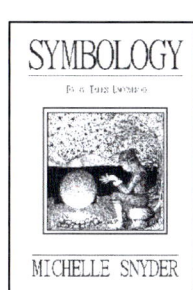

Fairytales Uncovered contains famous and obscure fairytales decoded by Symbologist Michelle Snyder and Doc E. Illustrated.

A Tale of Three Kingdoms series will take you on an unforgettable journey of magical beings, dangerous villains, and heroes with the courage and love to save the day. Reading level 13+

The first book in the trilogy *A Tale of Three Kingdoms, The Lost Unicorn* is an adventure with magic, love, jealousy, fairies, good and evil wizards, unicorns, a princess, and a hero. Shyla must find a mysterious hidden island to seek the help of a prince to save her kingdom.

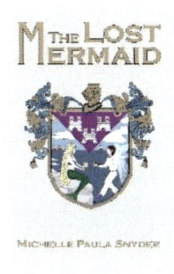

The second book in the trilogy *A Tale of Three Kingdoms, The Lost Mermaid,* in which a Mermaid Princess and a Fairy Prince get embroiled with a dark wizard and witch. The Prince must rescue the kidnapped princess and save both their kingdoms.

The third book in the trilogy *A Tale of Three Kingdoms, The Lost Dragon,* concludes a war between the Light Wizard and the Dark One, and the Pixie Queen is caught in the middle. She must seek the Dragons, who are only found in legend and myth.

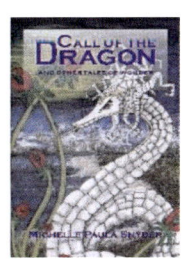

Call of the Dragon and Tears of the Dragon are collections of original short wonder tales for your enjoyment, enlightenment, and entertainment.

The Fairy Tales, Once Upon a Time Lessons, Classic: fairy tales within a fairy tale. This story is about Fair One who learns much about history, the moon, and life through much-loved fairy tales.

All proceeds benefit FREA

Websites and Blogs

FREA website: *www.freafoundation.space*

Enzmann Starship: *https://enzmannstarship.com*

The Symbologist: *www.whiteknightstudio.com*

Visit us on Facebook:

https://www.facebook.com/Enzmann Starship

Memberships are now available online. All *ENDEAVOR* publications and the *Enzmann Chronicles*, newly released science-fiction by Dr. Robert Duncan-Enzmann, are now in digital formats for multiple devices. Memberships help support the work of FREA and to publish the Enzmann Archive.

Tax-deductible donations can be made online at www.freafoundation.space, and by PayPal at *freafoundation @ gmail.com,* or by check to the address below.

FREA is supported in part by the generosity of Amazon customers who have chosen to use Amazon Smile and designated FREA as the non-profit they help with a small percentage of their Amazon purchases.

**Foundation for Research
of the Enzmann Archive, Inc.**

Foundation Headquarters

The Odlaw House, ca 1730

70 North Street, Grafton, MA, 01519

Telephone 508 839 4929

*Autumn 2020 on FREA's Odlaw House Gardens
Enzmann Archive Image*

www.freafoundation.space

Made in the USA
Middletown, DE
28 February 2021